Greenpeace in the Circumpolar North

This book explores Greenpeace's efforts to expand its engagement in the Circumpolar North in the 21st century and how this work is affected and informed by the organization's controversial legacy of anti-sealing campaigning in the 1970s and 1980s.

Presenting the fallout for peoples and cultures targeted by Greenpeace's anti-sealing campaigning as a pivotal dimension to the organization's history, the book argues that this history must first be acknowledged in order to understand how Greenpeace has developed more positive working relationships in some instances with northern Indigenous peoples in recent years. The book looks to dispel the misconception that Greenpeace is universally rejected in the Circumpolar North, whilst also highlighting that its engagement and alliances are being built in the shadow of its yet-to-be fully tackled history as a leading part of the anti-sealing movement.

Greenpeace in the Circumpolar North is ideal for courses and research with a focus on Arctic studies, environmental activism, and Indigenous studies, and for those interested in learning more about the complex legacy of Greenpeace.

Danita Catherine Burke is a senior research fellow based at the Center for War Studies and the international politics and border regions group at the Department of Political Science and Public Management, University of Southern Denmark. Dr. Burke was a Northern Scholars Visiting Research Fellow at the Institute for Advanced Studies in the Humanities at the University of Edinburgh while completing this book. Her recent publications include *Cultural Violence, Stigma and the Legacy of the Anti-Sealing Movement* (2023), *WWF and Arctic Environmentalism: Conservationism and the ENGO in the Circumpolar North* (2022), *Diplomacy and the Arctic Council* (2019), and *International Disputes and Cultural Ideas in the Canadian Arctic* (2018).

Routledge Explorations in Environmental Studies

Disruptive Innovations and the Environmental Crisis
Ethical, Practical, and Sociopolitical Concerns
Edited by Donald S. Maier, Justin Donhauser, and Michael Weber

Music and the Performing Arts in the Anthropocene
Nature, Materialities and Ecological Transformation
Edited by François Ribac, Isabelle Moindrot and Nicolas Donin

Working through Planetary Breakdown
Labour, Skill and the Changing Climate
Edited by Chantel Carr and Jesse Adams Stein

Politics of the Anthropocene and Climate Crisis in India
Seeking Socio-Ecological Transformations
Edited by Purendra Prasad and Lalatendu Keshari Das

Environmental and Technological Threats in the Arctic Region
Infrastructures, Geopolitics and Strategy
Edited by Mathieu Landriault, Magali Vullierme and Michael Delaunay

Impacts of Climate Change on Vernacular Landscapes and Cultural Heritage
Gül Aktürk

Greenpeace in the Circumpolar North
Lessons Learned from the Anti-Sealing Era
Danita Catherine Burke

For more information about this series, please visit: www.routledge.com/Routledge-Explorations-in-Environmental-Studies/book-series/REES

Greenpeace in the Circumpolar North

Lessons Learned from the Anti-Sealing Era

Danita Catherine Burke

Routledge
Taylor & Francis Group
LONDON AND NEW YORK

First published 2026
by Routledge
4 Park Square, Milton Park, Abingdon, Oxon OX14 4RN

and by Routledge
605 Third Avenue, New York, NY 10158

Routledge is an imprint of the Taylor & Francis Group, an informa business

British Library Cataloguing-in-Publication Data
A catalogue record for this book is available from the British Library

ISBN: 978-1-041-14352-9 (hbk)
ISBN: 978-1-041-14353-6 (pbk)
ISBN: 978-1-003-67402-3 (ebk)

DOI: 10.4324/9781003674023

Typeset in Times New Roman
by SPi Technologies India Pvt Ltd (Straive)

Contents

Acknowledgements

The research in *Greenpeace in the Circumpolar North* is compiled from work from projects that were funded by

- Joan Mitchell Travel Award, Wilfrid Laurier University
- Northern Scholars Visiting Research Fellowship of the Institute for Advanced Studies in the Humanities, University of Edinburgh
- J.R. Smallwood Foundation for Newfoundland and Labrador Studies
- Horizon 2020 Marie Skłodowska-Curie Actions fellowship

The book was completed while undertaking related research supported by the Nordic Council in Greenland's Nordic Arctic Programme.

Thank you to all who helped to make the research and writing of this book possible. This book includes archival research from the Laurier Archives and Special Collections (Wilfrid Laurier University) and the Center for Newfoundland Studies (Memorial University of Newfoundland and Labrador). Special thanks to the archivists who aided the research journey, especially thanks to Amanda Oliver (Laurier) and Colleen Field (CNS), who helped with identifying relevant files and facilitating access to them. The book also includes content from in-depth interviews with Greenpeace activists with experience campaigning on Arctic and northern issues. Thank you to the interviewees for participating in my research, and especially for consenting to the use of their names, as it permits a level of contextual detail about the interviewees and their work that makes for a richer, more interesting exploration of Greenpeace's evolving relationship with the Circumpolar North.

Thank you to my host institution, the Department of Political Science and Public Management, and the Center for War Studies, University of Southern Denmark (SDU), for your continuous support over the years. The completion of my research on this book would not have been possible without your willingness to aid me in my research journey. Thank you to my colleagues at SDU for their support, with special thanks to Anette

Schmidt, Ann Skovly, Kristina Bøhnke, Signe Pihl-Thingvad, Vincent Keating, Kira Vrist Rønn, and Ingvild Bode.

Thank you as well to the Institute of Advanced Studies in the Humanities (IASH), University of Edinburgh, and the Centre of Canadian Studies for supporting me as a Northern Scholars Visiting Research Fellow and for your flexibility as I completed my research, including this book, while navigating cancer treatment. Special thanks in particular to Ben Fletcher-Watson (IASH), Pauline Clark (IASH), Lesley McAra (IASH), and James Kennedy (Centre of Canadian Studies) for aiding in my access to university space and resources to complete the book project.

The completion of this book was delayed due to sick leave. Thank you to the doctors and nurses who were central in my cancer journey. And thank you to my friends for their time and support during treatment.

Lastly, thank you to my family, particularly my parents, for your support with everything. Without all their help, I would not be here to complete this book.

1 Introduction

One of the most famous, or infamous, environmental non-governmental organizations (ENGOs)[1] in the world is Greenpeace.[2] Formed in Vancouver, Canada 1971, the organization began as a small group of like-minded individuals working against nuclear testing in Amchitka, Alaska (Greenpeace New Zealand n.d.; Carter 2001, 136; also see Carter 2007). Greenpeace has since become one of the most recognizable environmental organization brands in the world.[3] With the North, and the Arctic in particular, acting as a bellwether for climate change (Burke 2022), Greenpeace has turned its attention to the Circumpolar North (Arctic and sub-Arctic) and is seeking to participate in regional protection and governance discussions and debates (Greenpeace USA 2025).[4]

Greenpeace's reputation in the Circumpolar North is complex and evolving.[5] This book argues that the organization, in ally-ship with various local stakeholders and rightsholders, is seeking to improve its ability to contribute to, and influence, regional decision-making. This in turn is motivating Greenpeace to try and improve its relationships with peoples in the Circumpolar North. However, Greenpeace's efforts at engagement in the North remain affected by the organization's legacy of decision-making and contribution to cultural violence and harm stemming in particular from its early campaigning during the anti-sealing protests in the 1970s and 1980s, and anti-fur activism more broadly (Burke 2023; also see Woods 1986; Kerr 2014; Burgwald 2016; Burke 2020).

Greenpeace in the Circumpolar North points to Greenpeace's legacy of anti-sealing campaigning as being a mixed blessing for the organization, focusing on its time as a trailblazer playing a leading rôle in the anti-sealing movement. On the one hand, Greenpeace's anti-sealing legacy is internally viewed by many Greenpeace members and supporters as a period of great expansion and solidification of the organization as a global player in environmental activism. On the other hand, this period is also when the organization and its members caused irreparable harm to many local peoples and cultures in the North; incidentally, many of those affected are

DOI: 10.4324/9781003674023-1

now important regional rightsholders and stakeholders that an organization like Greenpeace would benefit from positive relations with.

Greenpeace in the Circumpolar North discusses the long-lasting implications of Greenpeace's anti-sealing campaigning. During its campaigning, Greenpeace violated its principle of non-violence by contributing to and popularizing a wave of attitudes, behaviours, and ideas that used and compounded to cultural violence against vulnerable rural and coastal Indigenous and non-Indigenous fishers, their families, and communities. By failing to remain committed to principled non-violence, Greenpeace during this period self-inflicted significant long-term harm on its own working relationships with local rightsholders and stakeholders, most prominently, but not exclusively, with Inuit of Alaska, Canada, and Kalaallit Nunaat/Greenland. The self-inflicted reputational damage that impacts Greenpeace's ability to get traction for its contemporary efforts at regional engagement (Burke 2020, 2021a, 2024), though this legacy is not exclusively why some Indigenous peoples are wary of Greenpeace's outreach (for more, see Chapters 2 and 3).

This book is not intended to be a holistic overview of all of Greenpeace's projects or positions related to the North. *Greenpeace in the Circumpolar North* focuses on addressing the question: how is Greenpeace attempting to engage in the Circumpolar North/Arctic? The simple answer appears to be that Greenpeace is attempting to engage regionally through targeted messaging and outreach to Indigenous peoples and it hopes that through internal modifications and signalling that it has evolved from past missteps it can begin to build inroads to foster ally-ship on areas of mutual concern. *Greenpeace in the Circumpolar North* begins to unpack this answer in more detail.

Greenpeace in the Circumpolar North highlights Greenpeace's contemporary efforts at engaging on issues in the Circumpolar North and its focus on building connections with Indigenous rightsholders. Notably, Greenpeace in the 21st century has evolved considerably since 1971, in large part due to the work, skill, and insight of some individuals within the organization who have on-the-ground experience and personal cultural connections and awareness of issues directly impacting some peoples in the places where they seek to be involved. But Greenpeace as a whole is a brand that is still tied to and benefiting from the legacy of its earlier work, with its background in anti-sealing activism is an enduring part of the complexities of its northern engagement and organizational history, especially with Indigenous peoples.

Greenpeace's activities, attitudes, and prominence during the anti-sealing movement in the 1970s, in particular, linger in efforts to evolve and engage in some parts of the North. The legacy casts an inconsistent and complicated shadow for the organization and its ability to reconcile its values, notably its commitment to non-violence and to finding solutions for issues (Burke 2021a).[6] This shadow is likely due to a combination of the enduring hardship faced by local peoples as a result of the legacy of the anti-sealing movement and the anti-fur movement that is it connected to, coupled with the underpinning colonialism exhibited and normalized in the environmental and animal rights movements of that era. Greenpeace's anti-sealing legacy is also something for which many Greenpeace supporters are extremely proud. As such, to disown and denounce the anti-sealing period in the organization's history would risk damaging the base support that the organization has amongst its employees and broader support base, including its paid membership and donors (Burke 2020).

The Arctic, and the Circumpolar North more broadly, have grown in political and economic significance, being targeted by governments and corporations with big ambitions for economic development, especially in the non-renewable resource sectors (e.g. mining, and oil and natural gas exploration and extraction) (Sale and Potapov 2010; Käpylä and Mikkola 2016). As such, backing away from the Circumpolar North, and the Arctic in particular, is not an option for Greenpeace, but engagement with the region today cannot risk replicating the most egregious elements of its past mistakes, nor entirely disavow them either (Burke 2020, 2024). It is a complex situation that Greenpeace leadership is presently attempting to navigate, and as *Greenpeace in the Circumpolar North* highlights, it is a situation in which some inroads have been made.

The goal of this book is to provide some original insight into elements of Greenpeace's polar-focused work, taking particular notice of what are arguably its most successful attempts at regional engagement in recent years, to tease out the "why" behind these successes as part of a "lessons learned" reflection both for Greenpeace and activism in the region generally. *Greenpeace in the Circumpolar North* embeds its discussion of Greenpeace within a discussion about the philosophy of non-violence and Eurocentrism. This context is vital to understanding the geopolitical and cultural landscapes that Greenpeace is entering into in the Circumpolar North.[7] Therefore, to understand why Greenpeace has had some success in the Circumpolar North, readers need to understand the legacy of what is arguably its greatest failure: the controversial legacy of the anti-sealing movement and Greenpeace's pioneering role in elevating the movement's global reach and impact at the expense of vulnerable rural and coastal peoples and ways of life.

Structuring the Greenpeace Discussion: Audience Appeal

In 2022, I published the book *WWF and Arctic environmentalism: Conservation and the ENGO in the Circumpolar North* with Manchester University Press. The interviews with the World Wide Fund (WWF) for Nature and Arctic state and Indigenous representatives in that book occurred as part of the same project in which the Greenpeace interviews in *Greenpeace in the Circumpolar North* occurred. All the interviews are part of an exploration of the role of environmental and animal rights organizations in Arctic diplomacy. In the 2022 book, I developed a framework which is implied here in *Greenpeace in the Circumpolar North* to inform this research.

The 2022 book argues that WWF focuses much of its Arctic and northern engagement using an insider strategy to appeal to two primary audiences: "(1) Arctic governments and (2) northern communities and peoples, notably Indigenous peoples" (Burke 2022, 2). The way in which WWF, and all environmental and animal rights organizations, go about making themselves appealing to different audiences is grounded in four interrelated pillars: (1) legacy; (2) networks; (3) scientific engagement; and (4) communication style (Burke 2022). In contrast with WWF's twin-track audience approach, I argue in *Greenpeace in the Circumpolar North* that Greenpeace started campaigning using an outsider strategy, though it has shifted since the 1990s into a more hybrid insider-outsider approach (Carter 2001; also see Burke 2022, 25). It is now focused primarily on building/rebuilding relationships with Indigenous peoples and communities and looking to work with them on campaigns about the Arctic/North. However, the pillar framework I developed in *WWF and Arctic environmentalism* can help readers understand what Greenpeace is and why it may or may not appeal to certain audiences.

This pillar framework is grounded in the argument that if an audience has a generally positive perception of these four interrelated dimensions of an organization, then the organization has a relatively good or neutral reputation from the perception of that audience. Therefore, an organization may appeal to that audience in some manner if it would like to for some reason (e.g. seeking ally-ship, opening dialogue). Legacy represents an overall picture of an organization's work, its structure, and what it purports to stand for, and how it goes about its work. Networks point to the positive connections which an organization has at an individual level (members' networks) and at an organizational level, which it can call upon to help it achieve its goals and do its work. Scientific engagement means the ways in which an organization facilitates and interacts with scientific research and acknowledges and engages with Traditional Knowledge when "planning and [the] implementation of conservation efforts and strategic planning on environmental issues" (Burke 2022, 5). Lastly, communication style speaks to how an organization goes about doing its work and communicating about it to different audiences and "how that communication style has helped to characterize external expectations of the organization and how it is able to engage on certain topics with different actors" (Burke 2022, 5).

While all the pillars are interconnected, *Greenpeace in the Circumpolar North* emphasizes that the legacy of Greenpeace's anti-sealing activism, and to some extent its communication style during that campaign, are stumbling blocks for Greenpeace in parts of the North, as is unpacked in Chapter 3. Awareness of these issues within Greenpeace leadership goes back to the 1970s, though the bearing of its legacy in particular is more impactful on its audience appeal in some parts of the North versus others (Canada vs Russia).

But Greenpeace is a multifaceted actor with strengths and challenges. It has the capacity to appeal to different audiences, which varies depending on their relationship to the same series of features and facts about the organization. While Greenpeace has made Indigenous peoples their primary targeted regional audience, Indigenous peoples is an extremely broad category, on par with saying Europeans, for example. Reception to Greenpeace regionally amongst the many different Indigenous peoples, communities, and countries is not uniform, nor are their direct past and current experiences with the organization and its campaigns; a factor which must be considered when reflecting on the reception to Greenpeace by different Indigenous peoples and actors.

Greenpeace in the Circumpolar North discusses three cases in which Greenpeace has worked to foster good relationships with Indigenous peoples in the North. The book does not argue that these examples of positive relationship building are generalizable samples of Greenpeace's overarching relationship with Indigenous peoples. But they do help to illustrate that the organization's appeal, especially in terms of its networks and access to resources, helps facilitate work, making association with it appealing to some actors who may lack those tools to achieve their own objectives. There are also lessons that Greenpeace International and its offices, as well as other environmental and animal rights groups, can glean from Greenpeace's approach and self-reflections to date on the three cases explored in this book.

Depending on the extent to which the organization's other pillars – legacy and communication style (past and present) – also appeal, or at least not offend or contrast negatively with other actors, then the capacity to build relationships in some fashion may be possible with various audiences. This is why, I would argue, that Greenpeace has been slowly, though arguably inconsistently, working to address its legacy in the Circumpolar North. Some members in the organization have understood that to appeal to its desired target audience in the Circumpolar North – Indigenous peoples – Greenpeace's legacy, especially with regard to anti-sealing/anti-fur campaigning (though not exclusively) and what it represents (e.g. indicative of issues of classism and Eurocentrism in Western-based environmental activism) must be openly addressed.

However, as my previous research on Greenpeace illustrates, putting distance between the organization and aspects of its legacy is not easily done. While anti-sealing activism, for example, represents hurt and harm on the part of the organization towards Indigenous peoples and their rights, it simultaneously represents campaigning innovation, success, and growth to other equally, if not more important, audiences that Greenpeace must appeal more broadly to in order to continue, such as its core membership base and donors (Burke 2020).

Inspiration

This book builds on my research on Greenpeace and the legacy of environmental and animal rights campaigning in the Arctic and sub-Arctic. This work includes research on cultural violence and the history of the anti-sealing movement. My publications which inform this research include: "Nonviolence and Greenpeace: Apologies and Anti-Sealing Campaigning" (2024 – Arctic), *Cultural Violence, Stigma and the Legacy of the Anti-Sealing Movement* (2023 – Routledge), *WWF and Arctic Environmentalism: Conservationism and the ENGO in the Circumpolar North* (2022 – Manchester University Press), "The Case for a Greenpeace Apology to Newfoundland and Labrador" (2021a – The Northern Review), "Re-establishing Legitimacy after Stigmatization: Greenpeace in the North American North" (2020 – Polar Record) and "Bridging Troubled Waters: History as Political Opportunity Structure" (2014 – Journal of Civil Society) (Phelps Bondaroff and Burke 2014; Burke 2020, 2021a, 2022, 2023, 2024).

Throughout the chapters, there are long quotes from various interviews with Greenpeace members and Kangiqtugaapik/Clyde River[8] Mayor, Jerry Natanine (Burke 2021b). This is a deliberate writing choice, though it is a style that is not for everyone. This approach is taken for a number of reasons. First, I have a preference for letting research participants voice their own views in my research through quotes. In my experience, the use of long quotes can help minimize the intended meaning of an individual's perception or statements from getting obscured, context being inadvertently omitted or misrepresented, or potential misinterpretation seeping into source attribution. Using the words of interviewees directly, and providing context and challenge when needed, either in text or endnotes, such as some background on the questions asked that prompt certain replies, helps ground the information provided by interviewees. I also find that using direct quotes from interviews helps make it clearer to readers where certain ideas or information comes from. This, in turn, aids in making distinctions between my description and analysis and the views expressed by others.

Second, in my years researching and talking with individuals impacted by the anti-sealing movement, a key problem with how fishers/sealers and their communities were represented is the selective use of quotes and imagery to misrepresent local voices and experiences. Greenpeace is a culprit of this communication technique, as discussed in Chapter 3. It was important, therefore, in my critique and exploration of Greenpeace to minimize the replication of the same techniques used against sealers/fishers. This is attempted by using as many of the words of interviewees and providing context for what they say about the organization's past and current activities and attitudes, their personal roles and experiences, their reflections on relationships with rightsholders and stakeholders, and their opinions on campaigning outcomes, issues, and local factors affecting their work. The contextual information provides documentation for the readers of how Greenpeace members think about their work firsthand, and how they want others to see it, too, thereby hopefully making this primary data more accessible for future researchers working on this subject.

Third, many of the people I spoke to were campaigners on the ground, driving activities, and working on the issues explored by *Greenpeace in the Circumpolar North*. This puts these individuals in prime positions to reflect on events. The use of interviews, and long quotations from them, is especially useful in situations when setting the scene and describing outcomes, where research and sources may be limited or not as readily available in English. This does not mean that the reflections are all-encompassing or entirely truthful; to the best of my ability, I triangulate claims made by interviewees and sources and provide context and critique. But it helps to add new and relevant information and input into the conversation on Greenpeace's evolving interest in the North and Arctic that is either under-explored or largely missing, at least in English-based research sources.

Finally, in addition to helping this research, the quotes in *Greenpeace in the Circumpolar North* are important for current and future research on Greenpeace as a valuable source of data that can serve as part of Greenpeace's institutional memory. Greenpeace is an organization with a history of evolving its narratives about its past campaigns. In this sort of circumstance, data can be easily lost, manipulated, and rewritten if it is not documented before individuals retire, leave the organization, or die. *Greenpeace in the Circumpolar North* strives to be part of this documentation of Greenpeace's legacy in the North and Arctic, and quotes directly from the activists who did the work for *Greenpeace in the Circumpolar North* is a vital part of this documentation process.

Chapter Breakdown

Chapter 2 introduces Greenpeace. It provides readers with an introduction to the early days of Greenpeace, narrowing in on its formation and early works that have come to define the Greenpeace brand and highlight the complexity of Greenpeace's early work (Weyler 2004; Ahmed 2006; Fouhy 1995; Phelps Bondaroff and Burke 2014; Kalland 2009). It also introduces Greenpeace's problems during its anti-sealing campaigning (Patey 1990; Harter 2004; Burke 2021a, 2020). Chapters 2 and 3 both briefly discuss how the handling of the anti-sealing campaign in parallel to Greenpeace's internal struggles for ideological direction (Weyler 2004) eventually contributed to a partial public apology to Inuit, Indigenous, and coastal peoples in Canada for exhibiting and contributing to colonial thinking, behaviours, and attitudes while participating in anti-sealing protests (Kerr 2014).

Chapter 3, however, focuses on Greenpeace's anti-sealing legacy. It makes the case that the organization is at times pragmatic in its interruption of what nonviolence means and how it should affect what Greenpeace does and should make amends for. This chapter argues that it is through acknowledgement and reflection on Greenpeace's anti-sealing legacy that we can understand the efforts to improve, evolve, and grow the way that the organization approaches its northern-focused activism.

Chapters 4–6 argue that Greenpeace is focusing its attention on Indigenous engagement and highlight three standout cases of positive examples. The chapters focus on the work of some within Greenpeace to try and balance organizational ideals and ambitions while acknowledging problematic elements of its own, and the environmentalist movement's, legacy towards Indigenous peoples and others practising traditional lifestyles. Chapter 4 navigates Greenpeace's work with Sámi reindeer herders in Finland against the Finnish government and the logging industry (Nyyssönen 2022). Chapter 5 explores Greenpeace's ally-ship with the Inuit community of Kangiqtugaapik/Clyde River in Canada and its effort to stop seismic testing offshore of the community, which was approved as a result of the Government of Canada's failure to adequately adhere to consultation procedures and previously negotiated conditions for respecting Inuit rights (Supreme Court Judgments 2017; Tasker 2017). Finally, Chapter 6 examines Greenpeace's work in Northern Russia with local peoples, namely, Nenet and Khanty Indigenous peoples, to protect Numto, a nature reserve and holy area for local Indigenous peoples in western Siberia, from encroaching oil exploration interests (Greenpeace 2016) and ongoing problems stemming from Soviet-era oil industry infrastructure.

Greenpeace in the Circumpolar North seeks to demonstrate the work of select individuals and parts of Greenpeace who are attempting to navigate the space between organizational legacy preservation and expectations and the acknowledgement of past mistakes and push for change. This is not easy work to do with pressures coming from within Greenpeace as branches, members and donors around the world can hold different stances, desires, and levels of understanding about local situations, cultures, and economies in the North. But as *Greenpeace in the Circumpolar North* demonstrates, there is work ongoing, with some members of the organization attempting to evolve how they engage peoples and cultures on the ground in ways that are mutually beneficial and more tolerant and nuanced than previously exhibited. The anti-sealing legacy stands as a point of reflection in this internal evolution, not necessarily impacting all its northern engagement directly, but being a period in the organization's history from which lessons can be, and are being, learned, especially when it comes to engaging with Indigenous peoples in the North.

Notes

1 IENGOs are a subtype of the category of actors known as non-governmental organizations (NGOs). Michele M. Betsill and Elisabeth Corell define NGO as a "broad spectrum of actors from advocacy groups rooted in civil society to privately held multinational corporations and trade associations to research-oriented bodies that participate in international environmental negotiation processes using the tools of diplomacy" (Betsill and Corell 2008, 4) with the main characteristic of an NGOs being that they result from private initiatives and on principle act independently of governments (Betsill and Corell 2008, 12). NGO can be defined as "an organization that (1) is not formed by intergovernmental agreement, (2) has expertise or interests relevant to the international institution, and (3) expresses views that are independent of any national government" (Betsill and Corell 2008, 4). However, some scholars distinguish between NGO and international NGO, such as Bob Reinalda, who notes that

> NGOs are domestic actors when they confine their activities to their national political systems. They become transnational actors as soon as they operate across national boundaries, for instance by establishing a relationship with a similar NGO in another country. When various NGOs from three or more countries establish an international non-governmental organisation (INGO) to serve as a mechanism for co-operation among national NGOs in international affairs, NGOs through their INGO become international actors.
>
> (Reinalda 2001, 12)

2 Interview by Danita Catherine Burke with John Hocevar of Greenpeace USA, November 9, 2018; Greenpeace has a complex reputation and there is internal awareness of this. For example, Hocevar responded to the question "how do you think Greenpeace is perceived internationally?": "It really depends on who you ask. We are pretty controversial in our long history but at the same time we have gained a lot of respect from different actors at different levels in different sectors".

3 There is some margin of error when estimating the size of Greenpeace. For example, a 2010 report stated that Greenpeace had approximately 2.8 million supporters globally and is present in over 40 countries (Wang 2010). In 2018, Doyle reported that Greenpeace International

had "nearly 3 million supporters and offices in 41 countries" (Doyle 2018, 265). Roughly speaking, the organization appears to have, formally, at least 2.8–3 million supporters globally.

4 For example, Greenpeace USA (2025) states that "[c]limate change has already removed at least 75 percent of Arctic summer sea ice volume", thereby leaving the Arctic Ocean increasingly "exposed to exploitation and environmental destruction". They also argue for a Global Ocean Treaty to "protect the Arctic Ocean", which it presents as "under no form of protection" (Greenpeace USA 2025).
This is a misleading statement. While it is true that there is no encompassing Global Ocean Treaty or Global Arctic Ocean Treaty, much of the Arctic Ocean falling within the national jurisdiction of the five Arctic coastal states (Burke 2019), with Canada, for example, having the Arctic Waters Pollution Prevention Act in place since the 1970s (Burke 2017), there are other agreements that cover aspects of Arctic Ocean protection. The Arctic Council, for example, facilitated three legally binding agreements "enhancing international cooperation on issues related to maritime search and rescue, marine oil pollution, and Arctic scientific cooperation"; (1) Agreement on Cooperation on Aeronautical and Maritime Search and Rescue in the Arctic (as of 2011); (2) Agreement on Cooperation on Marine Oil Pollution Preparedness and Response in the Arctic (as of 2013); and (3) Agreement on Enhancing International Arctic Scientific Cooperation (as of 2017) (Arctic Council n.d., *International Cooperation*). The five Arctic coastal states (Canada, Kingdom of Denmark, Norway, Russian Federation and the United States) signed a legally binding fishing agreement to implement a temporary moratorium on fishing with the world's leading fishing nations (Iceland, China, Japan and South Korea) and the European Union to protect the Arctic Ocean from unregulated commercial fishing in the high seas portion of the Arctic Ocean (Burke 2019). The Agreement to Prevent Unregulated High Seas Fisheries in the Central Arctic Ocean came into effect in 2021 (Covey 2021).

5 There are no fixed, universally accepted definitions for the North and Arctic, either contemporarily or historically (e.g. Grant 1989; Elliot-Meisel 1998; Burke 2021a). However, what is broadly understood is that the North/Circumpolar North is a broader concept than the Arctic, and it can be subdivided into two categories that lack clear boundaries: the Arctic/High North and the sub-Arctic (Burke 2018, 6–8). There is debate on definitions for North, Arctic, and sub-Arctic, in part because of variations in national experiences which factor into how the areas are defined and depicted within societies and cultures, in combination with the fact that the concepts themselves are not static. Shelagh Grant captures the essence of the state of flux that the definition of the North has, when she stated in reference to the Canadian experience that " [t]he 'real' north keeps moving north, but never ceases to exist" with the north also being able to be "merely a 'state of mind,' directly related to one's own experience" (Grant 1989, 16; also see Elliot-Meisel 1998, 77).

6 Greenpeace underwent a rapid period of professionalization in the 1990s, moving away from its more radical origins and approach to activism, and became more coordinated across its various offices (Carter 2007). Mads Flarup Christensen reflected on this transitional period and the ways Greenpeace changed, and why, in his view. Christensen stated,

> The world has changed. We lived up through the 1970s and 80s in a situation where, globally speaking, the environment was not the centre of everyone's attention. If we were not along, then we were one of few that with our tactics, with our message, spoke about the environment. It was a marginal issue in many ways and therefore when Greenpeace did something, also given that we had a media landscape that was much more, at least in Europe and many other places, were very monopolized and very sort of that if you got through the evening news, then everyone knew what you were doing. It was very well suited for the very, very one-dimensional Greenpeace that basically saw a problem, sent activists out to put a camera on it, and delivered those cameras to an audience through the media that would then create political pressure. Very linear and we could replicate that again and again and again and achieve approximately the same result. I think what happened up through the 1990s was a complete change of the media landscape and a complete change of the topic of environment, climate change, etc. It came up and was the topic for so many organizations. There are so many actors and companies, renewable energy and a whole green industry track, and major political parties that took up the agenda as well. Which is a completely different landscape and I think that challenged Greenpeace to be much more solution oriented. It was no longer possible for us to just put the "think" on a problem and then expect someone else to solve it, right. We were expected to be much more closely aligned with the science, much more closely aligned with realistic solutions, and the understanding that the problems were also different.... There was a huge scientific debate about it. It's just different kind of problems, a different kind of media, a different kind of political landscape that just forced Greenpeace to be a completely different of organisation. If you then move further up toward the end of the 2008–2009 or even mid-2000s-ish, you are also in a situation where social media came about and you have the need to be able to talk much more directly with your audiences through those channels rather than though just mass media and traditional outlets. A lot of things changed that forced a professionalization, a specialisation and luckily, I think Greenpeace maintained its brand strength thereby we are able to attract supporters and thereby money and we could grow as an organisation....It moved from being an international organisation to being a global organisation and with staff numbers that probably went from 400–500 into the two, three, four thousand.
>
> (Interview by Danita Catherine Burke with Mads Flarup Christensen of Greenpeace Nordic, February 7, 2019)

7 *Greenpeace in the Circumpolar North* builds on my previous Routledge Focus publication, *Cultural Violence, Stigma and the Legacy of the Anti-Sealing Movement*, published in 2023.

8 Clyde River's name in Inuktitut has two common spellings: Kanngiqtugaapik and Kangitugaapik. Kangiqtugaapik is used in this book as it is the spelling used on the Municipality of Clyde River website.

References

Ahmed, Meena. 2006. *The Principles and Practice of Crisis Management: The Case of Brent Spar*. Palgrave.

Arctic Council. n.d. "International Cooperation in the Arctic". https://arctic-council.org/explore/work/cooperation/

Betsill, Michele M. and Elisabeth Corell. 2008. "Introduction to NGO Diplomacy." In *NGO Diplomacy: The Influence of Nongovernmental Organizations in International Environmental Negotiations*, edited by M. Michele Betsill and Elisabeth Corell. The MIT Press.

Burgwald, Jon. 2016. "Where Does Greenpeace Stand on Seal Hunting?" *Greenpeace*, 21 January. https://www.greenpeace.org/canada/en/story/424/where-does-greenpeace-stand-on-seal-hunting/

Burke, Danita Catherine. 2017. "Leading by Example: Canada and its Arctic Stewardship Role". *International Journal of Public Policy* 13(1/2): 36–56.

Burke, Danita Catherine. 2018. *International Disputes and Cultural Ideas in the Canadian Arctic: Arctic Sovereignty in the National Consciousness*. Palgrave Macmillan.

Burke, Danita Catherine. 2019. "Club Diplomacy in the Arctic." *Global Governance* 25(2019): 304–326.

Burke, Danita Catherine. 2020. "Re-establishing legitimacy after Stigmatization: Greenpeace in the North American North." *Polar Record* 56(e26): 1–12.

Burke, Danita Catherine. 2021a. "The Case for a Greenpeace Apology to Newfoundland and Labrador." *The Northern Review* 51(2021): 173–187.

Burke, Danita Catherine. 2021b. "The Relationship between Kanngiqtugaapik/Clyde River and Greenpeace: An Interview with Mayor Jerry Natanine." *Arctic* 74(1): 106–110.

Burke, Danita Catherine. 2022. *WWF and Arctic Environmentalism: Conservationism and the ENGO in the Circumpolar North*. Manchester University Press.

Burke, Danita Catherine. 2023. *Cultural Violence, Stigma and the Legacy of the Anti-Sealing Movement*. Routledge.

Burke, Danita Catherine. 2024. "Nonviolence and Greenpeace: Apologies and Anti-Sealing Campaigning." *Arctic* 77(3): 246–260.

Carter, Neil. 2001. *The Politics of the Environment*. Cambridge University Press.

Carter, Neil. 2007. *The Politics of the Environment: Ideas, Activism, Policy*. 2nd Edition. Cambridge University Press.

Covey, Nicole. 2021. "Legitimization of the Arctic Coastal States (A5) through the Central Arctic Ocean (CAO) Fisheries Agreement". *North American and Arctic Defence and Security Network (NAADSN)*. https://www.naadsn.ca/wp-content/uploads/2021/11/21-oct-Covey-CAOFA-PPrimer.pdf

Doyle, Timothy. 2018. "Environmental Non-Governmental Organisations". In *Handbook of Research on NGOs*, edited by Aynsley Kellow and Hannah Murphy-Gregory. Edward Elgar Publishing, 258–279.

Elliot-Meisel, Elizabeth B. 1998. *Arctic diplomacy: Canada and the United States in the Northwest Passage*. Peter Lang.

Fouhy, Ken. 1995. "Greenpeace Admits it Erred in Brent Spar Analysis." *Chemical Engineering* 102(10): 51.

Grant, Shelagh. 1989. "Myths of the North in the Canadian Ethos." *The Northern Review* 3/4(Summer/Winter): 15–41.

Greenpeace. 2016. "Local Activists Oppose Oil Industry in the Yamal District in Russia." July 20. https://media.greenpeace.org/archive/Local-Activists-Find-Broken-Pipeline-in-the-Yamal-District-in-Russia---Web-Video--Russian-Version--27MZIFJJ8MLUB.html

Greenpeace New Zealand. n.d. "The Birth of a Movement." https://www.greenpeace.org/new-zealand/about/our-history/

Greenpeace USA. 2025. "Protect the Arctic." https://www.greenpeace.org/usa/protect-the-arctic/

Harter, J. 2004. "Environmental Justice for Whom? Class, New Social Movements, and the Environment: A Case Study of Greenpeace Canada, 1971–2000." *Labour/Le Travail* 54(1): 83–119.

Kalland, A. 2009. *Unveiling the Whale: Discourses on Whales and Whaling*. Berghan Books.

Käpylä, J. and H. Mikkola. 2016. "The Promise of the Geoeconomic Arctic: A Critical Analysis." *Asia Europe Journal* 14(1): 203–220.

Kerr, Joanna. 2014. "Greenpeace Apology to Inuit for Impacts of Seal Campaign." *Greenpeace*, June 24. https://www.greenpeace.org/canada/en/story/5473/greenpeace-apology-to-inuit-for-impacts-of-seal-campaign/

Nyyssönen, Jukka. 2022. "Frame Alignment Between Environmentalists and the Sa'mi in the Forest Dispute in Inari, Finland Until the 2000s—Competing Conservation Needs and Obstacles for Co-Living With the Non-Human." *Frontiers in Conservation Science* 3: 925713. doi:10.3389/fcosc.2022.925713

Patey, Francis. 1990. *A Battle Lost: An Unsuccessful Attempt to Save the Seal Hunt*. Robinson-Blackmore Printing and Publishing Ltd.

Phelps Bondaroff, Teale N. and Danita Catherine Burke. 2014. "Bridging Troubled Waters: History as Political Opportunity Structure." *Journal of Civil Society* 10(2): 165–183.

Reinalda, Bob. 2001. "Private in Form, Public in Purpose: NGOs in International Relations Theory." In *Non-State Actors in International Relations*, edited by Bas Arts, Math Noortmann and Bob Reinalda. Ashgate.

Sale, R. and E. Potapov. 2010. *The Scamble for the Arctic: Ownership, Exploitation and Conflict in the Far North*. Frances Lincoln Limited Publishers.

Supreme Court Judgments. 2017. Clyde River (Hamlet) v. Petroleum Geo-Services Inc. Case number 36692. https://scc-csc.lexum.com/scc-csc/scc-csc/en/item/16743/index.do

Tasker, J.P. 2017. "Supreme Court Quashes Seismic Testing in Nunavut, but Gives Green Light to Enbridge Pipeline." *CBC News*, July 26. https://www.cbc.ca/news/politics/supreme-court-rulingindigenous-rights-1.4221698

Wang, L. 2010. "Greenpeace." In *International Encyclopedia of Civil Society*, edited by Helmut K. Anheier and Stefan Toepler. Springer. 810–811.

Weyler, R. 2004. *Greenpeace: How a Group of Ecologists, Journalists, and Visionaries Changed the World*. Raincoast Books.

Woods, S.J. 1986. "The Wolf at the Door: The Anti-Harvest Campaign Strikes at the Heart of North Aboriginal Economies." *Northern Perspectives* 14(2): 1–8.

2 The Founding of Greenpeace

Greenpeace started in 1971, first as the loosely formed group "Don't Make a Wave Committee", which came together to try and stop the US government from conducting nuclear testing in Amchitka, Alaska (Hunter 2004; Weyler 2004; Zelko 2013). Since then, the organization's brand and the scope of its work have grown exponentially. Now with millions of supporters worldwide (Doyle 2018), the organization has established itself as a major player in environmental activism. Timothy Doyle, in their work on environmental non-governmental organizations (ENGOs), states that Greenpeace is

> More likely to see environmental issues in post-material terms, as issues which cross local and national borders, affect[ing the] environment, and [in] need [of] a global approach… [T]he focus of Greenpeace is environmental awareness-raising at a global level, and direct action campaigns, which it hopes will lead to global policy change.
>
> (Doyle 2018, 265)

Scholarship on Greenpeace's engagement in the Circumpolar North needs expansion. Much of the present focus on Greenpeace's work in the Arctic or North is broad, focuses on Greenpeace's early work in the 1970s protesting American nuclear testing at Amchitka (Hunter 2004); Greenpeace's "Save the Arctic" campaign and work against oil exploration and extraction (Shapovalova 2010; Stewart 2015); and its anti-sealing legacy and its negative impact on Indigenous and non-Indigenous peoples (Morse and Zakrison 2010; Burke 2020, 2021a, 2021b, 2023). While this scholarship aids in improving our understanding of Greenpeace's northern-focused work, *Greenpeace in the Circumpolar North* introduces readers to some of Greenpeace's efforts to engage in the region in the 21st century, focusing on its contemporary efforts to engage with Indigenous peoples. In particular, Chapter 2 focuses on the foundation of Greenpeace and the central importance of the philosophy of nonviolence to both the organization's identity and how it reportedly formulates its activism.

DOI: 10.4324/9781003674023-2

The emphasis on the inception of Greenpeace and nonviolence is made in this chapter because, as Chapter 3 will further unpack, the early years of Greenpeace were fraught with internal conflict, which was played out during the organization's anti-sealing campaign. Greenpeace's anti-sealing campaign, and the anti-sealing and anti-fur movements in general, have had a profoundly negative impact on many Indigenous and non-Indigenous coastal and rural peoples and cultures in the Circumpolar North. This legacy is complex for Greenpeace to navigate as it seeks to engage northern audiences in contemporary times. *Greenpeace in the Circumpolar North* argues that it is through improving awareness of the connections between the organization's formation, its in-fighting during its anti-sealing activism, and its relationship with nonviolence philosophy that we can better understand why certain engagement (outlined in Chapters 4–6) have been relatively positively received for both the organization and the local peoples they have interacted with while campaigning.

Establishing Greenpeace

Greenpeace started in British Columbia, Canada (Pearce 1996, 73)[1] as the "Don't Make a Wave Committee" in 1969[2]; later becoming Greenpeace in 1971.[3] The Don't Make a Wave Committee formed to stop the US government from conducting nuclear testing plans at Amchitka, Alaska (Pearce 1996, 73); "The Amchitka voyage established the group's name in Canada. Greenpeace's next journey spread their reputation across the world" (Greenpeace International n.d., Our founders). During the anti-nuclear testing campaign, the committee transitioned into Greenpeace under the leadership of Bob Hunter (Greenpeace International n.d., Our founders; Greenpeace USA 2010a; Weyler 2021).

The name Greenpeace emerged after "Irving Stowe flashed the peace sign – as was his custom – and said 'Peace' [at the end of a meeting] …ecologist Bill Darnell made the off-hand reply: 'Make it a green peace'" (Greenpeace International 2007). The leadership and structure of Greenpeace in its early years were loose, and the organization was not a professional operation but rather a small collection of conservation and ecologically focused people like Bob Hunter, Marie Bohlen, Irving Stowe, Bill Darnell, Paul Watson, Patrick Moore, and David McTaggart (Greenpeace International n.d., Our founders; Greenpeace USA 2010a, 2010b).[4]

The initial activism of Greenpeace primarily targeted governments, such as the United States and the French governments, for their nuclear testing (Weyler 2004; Zelko 2013). In this way, the early Greenpeace brand was interwoven with the idea of a David versus Goliath fight; the relatively small, passionate, and new group of young environmentalists taking on the establishment to protect the planet.[5] The David versus Goliath imagery and what it represents are integral to the Greenpeace origin story and identity. Effort to portray Greenpeace, its members and causes as the David in situations was observable throughout the marketing of Greenpeace work in the 1970s–1980s, including against whale hunting; a campaign Greenpeace started in 1975 targeting commercial whale hunting particularly hunting by Soviet and Japanese whaling ships (Weyler 2004; Greenpeace USA 2010b) (discussed more in Chapter 3).

Weyler (2004) reported, for example, that Bob Hunter, a co-founder of Greenpeace and a leading voice in the early years of the "Save the Whales" campaign,[6] understood the resonance of the narrative and imagery of a small actor or cause as a focal point juxtaposed against observably larger, more powerful entities. Weyler stated that Hunter said: "We have to… show people that the whales are small compared to the technology of industrial whaling" (Weyler 2004). Hunter understood the juxtaposition could help convey the idea of vulnerability, and, in turn, help craft a message of moral rightness around the need to protect.

Though focused on commercial whale hunting, Greenpeace's campaigning negatively impacted traditional whaling, most especially whaling by Indigenous peoples, such as Inuit in Canada (Hopper 2014).[7] The lack of nuance demonstrated in its anti-whaling messaging also occurred in Greenpeace's anti-sealing campaigning, which was happening around the same time. Both campaigns cast large shadows over Greenpeace's relationship with many rural and coastal northerners. However, it is Greenpeace's first campaign against nuclear testing in Amchitka that is the most significant for the internalized Greenpeace identity and projection of itself.

Greenpeace's Amchitka voyage against nuclear testing solidified the organization's association of itself with the David role in campaigning. To this day, the organization's leadership markets the Amchitka campaign front and centre, desiring its members and supporters to reflect on it when thinking about the virtue and legitimacy of the organization's inception, as is evidenced throughout Greenpeace's various websites and the repetition of the origin story (e.g. Greenpeace Canada n.d.; Greenpeace UK n.d.; Greenpeace International 2007). Greenpeace International (2007) presents the Amchitka protests as "the founding voyage". The retelling of the Greenpeace Amchitka origin story (e.g. Hunter 2004) is an illustration of how Greenpeace frames itself as grassroots while weaving in its interpretation of its advocacy and relationship with sources of money together in a tale that also aims to demonstrate its elite connections within the entertainment world (Keziere 2023).

While the Amchitka origin story provides a compelling narrative about the motivations, methods, and early success of Greenpeace, interwoven in that story are key components to how Greenpeace has constructed its identity, operationalizes its campaigns, and ultimately aims to legitimize what it does to itself, its members and supporters, as well as external audiences. These standout components are (1) financial independence; (2) moral authority; (3) nonviolence; and (4) taking responsibility for actions.

Projecting financial independence is a big part of Greenpeace's modus operandi. Raising money and financing agendas is a major issue that almost all NGOs must deal with; Greenpeace is not unique in this respect. NGOs in general face serious decisions with regard to sources of funding, such as whether to accept government or corporate grants, donations, or partnerships, and if they do, which types of governments or corporate actors they determine to be acceptable and under what conditions. In large part, this tension between needing money and finding sources of it exists because money and its sources are intimately tied to an organization's credibility and independence (Berlie 2010; Burke 2022).

In the NGO world, taking money from governments or corporations is subject to debate about whether such a practice constitutes selling out (Doyle 2018). Some activists deplore taking money from government and corporate actors, especially as doing so has resulted in practices such as greenwashing. Greenwashing is a practice whereby an actor (e.g. a corporation) intentionally misleads "the public to believe that a company or other entity is doing more to protect the environment than it is", which distracts from and delays "concrete and credible action" (United Nations n.d.; also see Miller 2017; Bowen 2014).[8]

Ill-selected alliances or associations that infringe on NGO independence – real or perceived – are something to be avoided as much as possible; "It is ... fundamental that the NGO maintains its autonomy and its capacity to criticise, and it must not hesitate to criticize" (Berlie 2010, 37). Greenpeace is mindful of avoiding scenarios that infringe on its flexibility; as John Hocevar of Greenpeace USA noted when asked to reflect on the main strengths of Greenpeace, his response was first and foremost "our independence".[9]

Taking money from an actor that would otherwise be subject to an NGO's criticism is a prime way for an NGO to undermine itself. Berlie (2010, 35–36) notes that the "capacity of confrontation is one of the resources an NGO possesses and often uses against firms...an alliance would therefore be a means of preventing the application of this capacity". Accepting corporate money enables the donor a level of control over an NGO, undermining it by neutralizing its capacity for confronting that same donor. NGOs can, therefore, find themselves in situations where their reputations and credibility appear to be purchased through a donation to an actor in exchange for a patina of green credentials through an association with an environmental NGO without doing substantial work or making changes to match their implied new image. For example,

> In 2007, the Sierra Club partnered with The Clorox Co. to promote new eco-friendly products that Clorox developed, by permitting the

> company to use the Sierra Club logo on its products in return for a profit share in the venture. But because Clorox' s brand is synonymous with its signature product, chlorine bleach, the decision by the Sierra Club executives to partner with Clorox caused internal fracturing within the organization... The fallout of the partnership illustrates ... taking corporate money can push an organization to the point where its core supporters believe the IENGO's credibility is compromised.
>
> (Burke 2022, 62–63)

The Sierra Club's experience with Clorox demonstrates that sources of money can have long-running implications for an NGO and its reputation.

Greenpeace leadership is aware of the complexities and implications of sources of money and types of allies, and has been aware since the organization's inception. Conscious of the optics and the liabilities of alliances, Greenpeace is selective about its connections.[10] This is why, for example, Greenpeace stresses that it was "the concert" that got the organization financially up and running; a detail that Greenpeace makes a deliberate point of reiterating in its communication about its history.

> We had our name, but it was soon clear that selling 25-cent buttons wouldn't bring in the cash needed to buy a boat. Someone had the idea to put up a rock concert. A few phone calls later, Joni Mitchell said she would be playing. Chilliwack and Phil Ochs confirmed, before Joni called again to say she would bring a special guest: James Taylor. None of them wanted any money for the night....The sixteen thousand that filled Vancouver's Pacific Coliseum left the concert entranced. Afterwards, attendance at the meeting swelled, the money started to pour in. By the end of October, the group had raised more than $23,000. Greenpeace was ready to go.
>
> (Greenpeace International n.d., Our founder)

From its beginning, Greenpeace's own telling of its founding story communicates clearly as a subtext that leadership today understands which is that money complicates organizational operations. Greenpeace, therefore, seeks to highlight a long-standing financial legacy linked to grassroots fundraising to portray a sense of continuity with financial independence and sources of money that extend to the present times.[11] At the same time, the narrative conveys that the group's early leaders were aware of the role that sources of money, and perceptions of those sources, could have on its ability to do its campaigning and its flexibility and independence.

At the same time, the Greenpeace re-telling of the Amchitka voyage reveals the self-belief that Greenpeace founders had that they were "saving the world through our moral example" (Greenpeace International n.d., Our founder). Moral legitimacy is extremely important for NGOs (Walton et al. 2016) as it is "the most meaningful type" of legitimacy for NGOs (Baur and Palazzo 2011, 584) and is assessed by evaluating a NGOs' "outputs and consequences (doing the rights things), techniques and procedures (doing things right), categories and structures (the right organisation for the job), and leaders and representatives (the right person in charge of the tasks)" (Liu et al. 2013, 635).

Greenpeace's version of the Amchitka voyage also discloses infighting during the journey (Greenpeace International n.d., Our founder). Greenpeace's positioning of itself and self-identification as a moral example to the public while admitting that in reality "we spent most of our time at each other's throats" (Greenpeace International n.d., Our founder; also see Greenpeace International 2020), is interesting and important for understanding the progression of Greenpeace after this founding narrative and the events in the subsequent years during its anti-sealing protests (Burke 2021a).

That said, Greenpeace's claim to moral authority cannot be considered in isolation; it is tied closely to the organization's projection of itself as a nonviolent organization embedded in the tradition of principled nonviolence akin to the work of Gandhi. The level of importance of the nonviolence philosophy and the complexities of it in activism and for activists who claim adherence to it are elaborated on more fully in the next section. At this stage, however, it is important to note that the philosophy of nonviolence is tied to pacifism for Greenpeace and its earliest representation of itself and its approach to activism; "The group organised a boat, the Phyllis Cormack, and set sail to Amchitka to 'bear witness' (a Quaker tradition of silent protest) to the nuclear test" which was seen as "novel and non-violent, if not entirely non-offensive, way of showing" opposition to the nuclear testing (Greenpeace International n.d., Our founders). The self-belief of the organization that its work is, and has always been, nonviolent from its inception is very important because it ties into the moral authority that the organization, and by extension its activists and supporters, argue that they have on the topics that they speak about and do activism on in the public arena.

Relatedly, a key part of Greenpeace's identity is the idea that the organization and its members always take full responsibility for their actions; a concept tied to the organization's commitment to nonviolence and moral authority.

> Those taking part in non-violent direct actions show their respect for the rule of law by taking full responsibility for their actions. We can only take action because ordinary individuals make a personal choice to help expose an environmental crime or stop environmental destruction and are willing to accept the consequences of their actions.
>
> (Greenpeace Canada n.d., Protesting peacefully)

The promotion of Greenpeace as a responsible organization owning its decision-making and the conduct of its members is tied to the prominent identity elements displayed in the origin story; being very conscious of sources of money and their implications on organization independence and ability to present their work as morally informed by nonviolence philosophy.[12]

For Greenpeace, nonviolence is ingrained in its identity and marketing of its way of operating on the world stage.[13] However, questions have been raised in academic scholarship about what nonviolence is (Randle 1994; Sharp 2005; Atack 2012; Ackerman and Kruegler 1994) and whether Greenpeace is actually nonviolent, especially in the context of its anti-sealing campaigning (Allen 1979; Harter 2004; Dauvergne and Neville 2011; Burke 2020, 2021a, 2024). The definition of nonviolence is sometimes used to mean resistance "without using or threatening physical harm against the opponent" (Chenoweth and Cunningham 2013, 72; also see Baldoli and Radaelu 2019).

The literature on nonviolence subdivides it into two broad groups: principled and pragmatic (Bharadwaj 1998; Sharp 2005; Atack 2012). The distinctions between the forms of nonviolence are significant. Iconic grassroots activist figures like Mahatma Gandhi and Martin Luther King (MLK) are associated with nonviolence, in particular, principled nonviolence (Bharadwaj 1998; Sharp 2005; Atack 2012).[14] However, the "nonviolence [activism] around the world … [is] being carried out largely in the pragmatic mode … [and] the majority of followers are attracted mainly to the charisma and effectiveness of leaders and not to their commitment to principled nonviolence" (Bharadwaj 1998, 79).

The reality is that "the vast majority of the instances of nonviolent political action or civil resistance involve individuals or groups who are using it for pragmatic reasons" (Atack 2012, 9). The desire of actors to present themselves as adhering to principled nonviolence is tempting with many upsides. The use of the term nonviolence does not make these distinctions readily evident to consumers of activist content, so the general public is not likely to recognize that the term nonviolence is not straightforward. As a result, applications of nonviolence are "bound to lead to an opportunistic or half-hearted acceptance of nonviolence, with the result that the temptation to use other means will lead to the eventual corruption or destruction of the good end" (Bharadwaj 1998, 80).

Ultimately, there is status to be gained and leveraged through the association of an organization and its members with established figures and causes in society that are imbued with moral authority like such as Gandhi and Dr. King. NGOs that claim to adhere to nonviolence may seek to associate themselves with such established figures to try and inhabit and inherit the moral legitimacy (Liu et al. 2013) these actors possess in our society and the public consciousness (Baur and Palazzo 2011, 584; Marberg et al. 2016, 2737–2738).[15]

For Greenpeace, the idea that the organization has always been connected to the philosophy of nonviolence is paramount to who the organization, its employees, and supporters think they are as "Greenpeace". This is stressed on the Greenpeace UK website, for example, which notes that their direct action is not just nonviolent but peaceful: "[t]aking peaceful, non-violent direct action has always been at the heart of Greenpeace's work" since it was founded in 1971 (Greenpeace UK n.d.).[16] Backing up the claims that the use of nonviolence is core to Greenpeace, Greenpeace Canada stresses that "[t]hose taking part in non-violent direct actions show their respect for the rule of law by **taking full responsibility** for their actions" (Bold text in the original, Greenpeace Canada 2022).

Therefore, the centrality of nonviolence in the organization's identity is evident, as Greenpeace's origin story is about the bringing together of green thought and peacefulness (Greenpeace UK n.d.).[17] But balancing green thought and peacefulness has been a challenge for Greenpeace. For example, Paul Watson was one of the early leaders in Greenpeace in the 1970s. Watson is a charismatic, but controversial, leader (see Bharadwaj 1998, 79),[18] and leaders within Greenpeace struggled with Watson's personal interpretation of nonviolence, which resulted in questionable events during Greenpeace's direct action on the ice protesting seal hunting in 1977 (more on this in Chapter 3). Rex Weyler, a member of Greenpeace during that era, observed that "Watson tended to push the end of nonviolence" and undermined Greenpeace's representation of itself as nonviolent at that time (Weyler 2004, 457). Ultimately, it was Watson's interpretation of nonviolence and how he went about teaching and putting it into action that became the final straw which led to Greenpeace's leadership voting him out of the organization (Weyler 2004; Burke 2021b).

To this day, Greenpeace's adherence to the philosophy of nonviolence is emphasized by its members. Sune Scheller of Greenpeace Nordic (Denmark) reflected, "you know it's like you have Greenpeace and its two words, the green part and the peace part. The peace part is a more social, society wide thing, whereas the green part is the environmental part".[19] Sini Harkki of Greenpeace Nordic (Finland) also noted the importance of nonviolence, which members spoke to, appearing to be closely identified with the peace dimension of the organization's name:

> What makes us a little bit different from the traditional nature conservation movement that has this colonial background is that we've had "peace" as the second part of our name from the start. I think being against wars and for human rights has been in our DNA from the start.[20]

And yet, it was also observed that the green side of the organization has a tendency to dominate the peace element. In Scheller's opinion: "I think the peace part is maybe not taking up as much space in Greenpeace as it has done in the past".[21] Harkki also expressed that:

> I think the early 2000s we definitely did less of the peace element and it maybe had natural reasons because we work so much on disarmament and nuclear weapons and some steps were taken and some of the most acute threats were no longer there. Then it was maybe natural to down-prioritize that. But then I think that when strong currents against civil rights arose from different parts of the planet it became clear that we needed to strength that side again [the peace side].[22]

The tension and battle for dominance between the green and peace parts of the Greenpeace identity go back to the early years of the organization, as will be discussed more in Chapter 3.

In sum, Greenpeace's brand is strongly associated with the philosophy of nonviolence. The organization repeatedly weaves in nonviolent activism into its narrative about its foundation, its way of operating, and its perception of taking accountability for its actions. However, while the term nonviolence is used extensively in Greenpeace's communication about itself, as Chapter 3 will elaborate, interpretations of nonviolence and the application of those interpretations have created a complicated legacy for Greenpeace in the North.

Environmentalism and Colonialism

Greenpeace does not operate in a vacuum. To some extent, the organization's efforts to carve out space for itself in environmental discussions involving the Arctic and North are impacted by wider issues of colonialism and Eurocentrism in the environmental movement. This means that Greenpeace's foundation and legacy are informed by how it is and has been influenced by attitudes, actions, and beliefs in environmental activism more broadly. Eurocentrism and colonialism are important factors that are part of this discussion, and should be acknowledged when trying to get a sense of challenges associated with Greenpeace's legacy, its approach towards its anti-sealing legacy (discussed more in Chapter 3), and efforts to engage Indigenous audiences in the Circumpolar North.

Eurocentrism is an offshoot of ethnocentrism. Ethnocentrism is "the regard of one's own ethnic group or society as superior to others" (Alatas 2007, 1491). Eurocentrism is a term that emerged in the 19th century (Alatas 2007) and was popularized in the 1980s (Wintle 2021). Basically, Eurocentrism means either (a) the belief that Western society is superior to other parts of the world or (b) the tendency to set Western experiences as the default norm from which the other parts of the world should be compared and judged (Brolsma et al. 2019, 12). The term reflects the

"European, and generally western concerns, culture, politics, and values are focused upon to the exclusion of others…and cannot be dissociated from the political, economic, and cultural domination of Europe and, later, United States" (Alatas 2007, 1491). Overall, though Eurocentrism has its origins in Europe, "it is not confined to Europeans or those in the West" and is based on a basic fallacy that European culture is universal (Alatas 2007, 1491).

Eurocentrism is rooted in colonialism. Colonialism is associated with state-based politics, but in recent years, the practices, approaches, and mindset that typify colonialism are now recognized to be part of the environmental movement and activism that emerged prominently in the 20th century as "colonialism dispossesses land not just through removal" (Erickson 2020, 112). Bruce Erickson in their work on the links between colonialism and environmentalism states that "[e]cological protection has come to shape the political life of colonialism" with an attempt in Canada at least "to place environmental collapse and change as the defining problem of all of humanity in the future to come… [which] legitimizes the continual colonial assertions of jurisdiction through conservation"; a framing and push by environmentalists that is being resisted by Indigenous peoples (Erickson 2020, 112).

Overall, there is a well-argued connection between environmentalism and Eurocentrism, or as Blaut (1999, 391) referred to it: "long-standing and happy marriage between environmentalism and Eurocentrism" (also see Erickson 2020). This happy marriage is observable in the way that many Western and urban-based environmental activists and their supporters, for example, frequently romanticize nature and erase or downplay the right to independent decision-making of the local peoples in areas, like the Arctic and North, they want to impact. As a result, Indigenous peoples, and often working-class or lower socio-economic stratified people, have been sidelined or undermined by environmental and animal rights activists who view them as obstacles. As Kashwan and Hasnain (2024, 25) note, mainstream environmentalism often fails to hear the voices of "racial minorities and the poor".

The undermining of racial minorities and the poor and presenting, and treating, them as a problem for nature is at the heart of the complex legacy of environmentalism, the anti-sealing movement, and the Greenpeace campaign within this atmosphere of problematizing the coastal peoples that have lived and worked with seals. Greenpeace's anti-sealing campaign, and the wider movement, exhibited textbook colonial attitudes, actions, and beliefs, as the organization conceded in a 2014 apology to Inuit of Canada, Indigenous and coastal peoples (Kerr 2014). But Greenpeace's anti-sealing work was central in making the organization a household name and international powerhouse in the environmental movement, with many of Greenpeace's donors and supporters very proud of this period in the organization's history (Burke 2020). So to some extent, the organization is stuck, needing and, to some extent, wanting distance from the broader baggage of colonialism and Eurocentrism that its anti-sealing campaign legacy typifies, but also knowing that the campaign helped make the organization's brand. To completely disavow its work could be tantamount to brand suicide for Greenpeace.

Conclusion

Greenpeace started in 1971 with a clear focus as a single-subject interest group against nuclear testing. Its early leaders had a keen awareness of the complications that accepting money from certain sources could have on its campaigning flexibility and took steps to minimize compromising its independence by limiting avenues for income. As will be discussed in Chapter 3, Greenpeace's promotion of itself as principled nonviolent is part of the reason why the anti-sealing movement is such an aberration for the organization's legacy and how it seeks to learn and evolve in its approach to Circumpolar North engagement in more contemporary times (see Chapters 4–6).

For an organization to claim genuine commitment to nonviolence from a principled perspective, it must illustrate that commitment through its actions and not just words. While this point holds true for pragmatic nonviolence as well, proponents of pragmatic nonviolence are more oriented towards the usefulness of the strategy and tend to focus on the success and power that such a strategy may gain for them. Greenpeace is an organization that very clearly aims to associate itself with principled nonviolence, linking itself to the moral legitimacy and legacy of prominent figures of principled nonviolent action such as Gandhi. However, deep in its history, in the early days of its campaigning against seal hunting, leaders in the organization made decisions that violated how they claim to think about nonviolence, and people were harmed as a result. The organization has subsequently supported those early decisions by remaining silent about those actions committed by its members, though, as Chapters 3 and 4 will discuss more, initial work towards reconciliation for elements of this past has started.

Notes

1 There are small divergences in the retelling of many aspects of the organization's early history. For example, Greenpeace USA (2010a), in a 2010 statement, claims the organization was formed in 1970 by Phil Cote, Irving Stowe, and Jim Bohlen. According to Greenpeace International, Cote, Stowe, and Bohlen are the first directors of the "Don't Make a Wave Committee", and Stowe and Bohlen are part a

the group of founding members of the committee, with Cote acknowledged as "a student" at that time (Greenpeace International n.d., Our founders). Greenpeace International's website states that "there was no single founder", though it does highlight Bob Hunter, David McTaggart, and Dorothy and Irving Stowe as key founding members. Dorothy and Irving Stowe, Marie and Jim Bohlen, Ben and Dorothy Metcalfe, and Bob Hunter are credited for the establishment of Greenpeace's predecessor, "Don't Make a Wave Committee" (Greenpeace International n.d., Our founders).

2 While 1969 is the year stated most often as the year that the "Don't Make a Wave Committee" formed (Pearce 1996), on the Greenpeace International website, it also states 1970 as the founding year of the "Don't Make a Wave Committee" (Greenpeace International n.d., Our founders).

3 Greenpeace was part of the second wave of environmentalism, which emerged starting in the 1960s (Carter 2007). As major environmental disasters were occurring and the public consciousness about the ecological impacts of war, including the threat of nuclear war and nuclear testing, rose, environmental issues began to get more public attention (Kashwan and Hasnain 2024). The "Crying Indian" public service announcement (PSA) that ran starting in 1971 (the same year as Greenpeace's founding) and into the 1980s in the United States in particular, appears to have been highly effective in shaping a public ideal about the environment. In the advertisement:

> Iron Eyes Cody, an actor in Native American garb, paddles a birch bark canoe on water that seems, at first, tranquil and pristine, but that becomes increasingly polluted along his journey…As the lone Indian ponders the polluted landscape, a passenger hurls a paper bag out a car window. The bag bursts on the ground, scattering fast-food wrappers all over the Indian's beaded moccasins. In a stern voice, the narrator comments: "Some people have a deep, abiding respect for the natural beauty that was once this country. And some people don't." The camera zooms in on Iron Eyes Cody's face to reveal a single tear falling, ever so slowly, down his cheek.
>
> (Kashwan and Hasnain 2024, 5)

The commercial typifies how Indigenous peoples have been stereotypically viewed by environmental activists and their supporters from this era, and the enduring power of the imagery. Ironically, however, Cody is not Indigenous (Paradis 2023); Cody is what is commonly referred to today as a "pretendian" (Kolopenuk 2023). According to the Indigenous news outlet ATPN, Cody claimed that his father was Cherokee, but it was later discovered in the late 1990s that, in fact, he was born to immigrants from Sicily and he was baptised Espera Oscar de Corti (Paradis 2023). In a further twist to the story, Kashwan and Hasnain (2024, 5) report that the "PSA was sponsored by beverage companies to undermine environmental groups that were seeking to hold industry groups responsible for the proliferation of disposable items that triggered the solid waste crisis".

4 Watson was expelled from Greenpeace in 1977 (more on this in Chapter 3). Moore remained with the organization in the subsequent years, but left in 1986. Both Watson and Moore left Greenpeace on bad terms with the organization (Moore 2010; Essemlali and Watson 2013), and in 2010, Greenpeace USA felt the need to release a statement on Patrick Moore denouncing his continued representation of himself as a founding member of Greenpeace. Greenpeace USA (2010a, 2010b) stated,

> Patrick Moore often misrepresents himself in the media as an environmental "expert" or even an "environmentalist," while offering anti-environmental opinions on a wide range of issues and taking a distinctly anti-environmental stance. He also exploits long-gone ties with Greenpeace to sell himself as a speaker and pro-corporate spokesperson, usually taking positions that Greenpeace opposes… Patrick Moore frequently portrays himself as a founder or co-founder of Greenpeace, and many news outlets have repeated this characterization. Although Mr. Moore played a significant role in Greenpeace Canada for several years, he did not found Greenpeace.

According to Patrick Moore, his departure from Greenpeace in 1986 centres on diverging views about the use of scientific data in campaigning, specifically a falling out on the science about chlorine and Greenpeace's support for a global ban on chlorine, which Moore argued was poorly informed and not in the best interest of global health. Moore has a PhD in ecology, and his education-based expertise was in contrast to other leaders in the organization in the 1980s, he reflects, "after six years as one of five directors of Greenpeace International, I observed that none of my fellow directors had any formal science education. They were either political activists or environmental entrepreneurs" (Moore 2008).

5 The use of narratives and orchestration of situations designed to echo David versus Goliath dynamics is a powerful, and well-established, path in environmental activism. David versus Goliath stories are very marketable and lean into parallels with the biblical story with "heroic activists bolding taking on high-handed governments and greedy corporations" (Mauch et al. 2006, 3; also see Santana 1996; Cianchi 2015).

6 Greenpeace started to work against whale hunting in 1975.

7 In 2014, Greenpeace Canada expressed "honour" in support of Inuit in Canada from the community of Kangiqtugaapik/Clyde River when community members killed a pregnant bowhead whale in "the community's first whale hunt in more than 100 years" (Hopper 2014). The Canadian newspaper, National Post, reported that Greenpeace Arctic campaigner Farrah Khan stated in response to the questions about Greenpeace's newfound open support for Inuit whaling in Canada that "Greenpeace respects the rights of Clyde River and other indigenous

communities to sustainable, traditional hunting and fishing"; a position on whale hunting that has left some Greenpeace supporters puzzled as Greenpeace's "name is synonymous with Save the Whales" (Hopper 2014), a campaign it had launched in 1975 that "eventually helped secure an international ban on commercial whaling" (Greenpeace USA 2010b).

8 Greenwashing has become a growing trend and problem in efforts to address climate change and the climate crisis. Miller (2017) argues that "cultural institutions endeavour to legitimize the harm they cause by promoting themselves as good environmental citizens". However, the term greenwashing has come under criticism. Bowen (2014, 15), in their work on greenwashing, and what they are is the wider concept of symbolic corporate environmentalism, argues that

> The academic decision of the term "greenwashing" has evolved in the past twenty years and has recently become more specific to deliberate, disclosure-based form activities. In contrast, broader symbolic corporate environmentalism resonates after greenwashing in the sense that it builds on an earlier, more politicised and critical version of greenwashing from twenty years ago.

9 Interview by Danita Catherine Burke with John Hocevar of Greenpeace USA, November 9, 2018.

10 Interview by Danita Catherine Burke with John Hocevar of Greenpeace USA, November 9, 2018. For example, in the interview with Hocevar, they expressed in response to the question "what kinds of actors approach you [Greenpeace USA]?":

> It could be everyone from politicians to NGOs and even to corporations. Also at times TV productions. You know Netflix could be working on a series and ask us for advice and access to our footage or the New York Times could be [working on a piece] and ask us for advice and access to ships...Those are all real [examples of Greenpeace USA being approached].

In a follow-up question, "Which of these actors are you most interested in partnering with and why?", Hocevar replied,

> We know that we need to, that education is important so as far as communication, if they have a large reach and if they have a reach in and for strategic audience, if they have more credibility than less, they are more serious, all of those things would influence our decisions. Similar with NGOs, are they [serious], do they have access that complements ours. Are we bringing things to the table that support each other rather than reinventing the wheel or competing?

11 Greenpeace's deliberate effort since its inception to make its fundraising grassroots and to limit its sources of money continues into its contemporary work. Doyle (2018, 265) states that "Unlike many other transnational ENGOS, Greenpeace International refuses to accept corporate or government funding", whereas some organizations "such as the WWF [World Wide Fund for Nature], are heavily funded both by nation states and by large business interests".

12 Greenpeace moved towards professionalization in the 1990s, and the implications of this move, especially post the Brent Spar fallout and Greenpeace's evolving relationship with the use of science and scientific research to inform its campaigning (Fouhy 1995, 51; Knott 1995, 25; Ahmed 2006). In 1995, Greenpeace was forced to publicly admit that it got its scientific analysis on potential toxic waste from tanks of the Shell Oil Brent Spar oil platform wrong when it protested against the platform's offshore disposal (Fouhy 1995, 51). Greenpeace also strategically started to shift its reputation away from being an outsider and rebel, as it had depicted during its anti-sealing activism days, towards a more hybrid model of advocacy to potentially broaden its reach (Carter 2007). When discussing NGOs, the insider strategy is characterised by "seeking to attain influence by working closely with negotiators and governments by providing policy solutions and expert advice. They also engage in knowledge construction and the production of research-based reports and papers on particular topics" (Gulbrandsen and Andresen 2004, 56). An outsider strategic approach is characterised by activists avoiding or severely limiting the extent to which they will work with actors like governments and big business or take funding from these types of actors in order to preserve autonomy while undertaking activities such as protests and campaigns that use tactics like deliberating causing meeting disruptions and direct action, like interfering in activities such as chaining oneself to trees and vehicles to stop logging or other industry activities in order to call for and motivate changes (Fogarty 2011, 210).

13 Presently, Greenpeace works to distinguish what nonviolence means for the organization by building in training for its activists. For example, Sune Scheller noted in his interview that Greenpeace Nordic (Denmark) recognized the necessity of ensuring that its activists are trained,

> To make sure that you understand the core values of Greenpeace, nonviolent direct action, always peaceful of course, what's the story about civil disobedience and what is our approach to it, how do I behave myself being faced with authorities, police to make sure that I'm not alone, not violent against these people but also so that I don't end up being perceived as violent – rightfully or wrongly.

Scheller argued that this training is important because "It's to make sure that people…understand because nonviolence is so important to us and we are being accused every time we do something of being violent. Half the population in Denmark thinks to some extent we're violent". When I followed up by asking why some people in Denmark have this perception of Greenpeace, Scheller replied,

> I think it comes from the understanding of violence. So for example, in some peoples' eyes trespassing is perceived as a violent act. It's a violation of private property but it's not violent as such in our sense, as in physical. Anyways that's something that we strongly battle with.
>
> (Interview by Danita Catherine Burke with Sune Scheller of Greenpeace Nordic, September 3, 2018)

Interestingly, in Scheller's reflections, the definition of nonviolence appeared to mean non-physical violence. However, when I pushed Scheller and asked, "Just to be clear, when you mean nonviolent you mean purely a physical violence? You don't take into account psychological or mental abuse into this?" they stated, "I think that [psychological or mental abuse] should be included, for sure".

14 Atack (2012, 9) states that

> Pragmatic proponents of nonviolent political action use these methods, because they seem to be the most effective under the circumstances or because armed resistance or armed conflict are not realistic options. Principled proponents of nonviolence, on the other hand, attach some deeper ethical or religious significance to avoiding or replacing violence in political action and social relationships more generally.

15 NGOs are tied to the concept of moral legitimacy because they generally make claims to be operating on behalf of the common good.

16 In conversation with Greenpeace members in the research for this book in 2019–2020, it was noted that Greenpeace UK is relevant in a discussion of Greenpeace's work in the North, as there is no Greenpeace office in Iceland, and work that the organization does that pertains to Iceland often involves the Greenpeace UK office.

17 Baldoli and Radaelu (2019, 1166) argue that the hyphen is significant in the nonviolence literature. In the literature, the hyphen is often viewed as a demarcation of mindset towards the concept of nonviolence. Those that tend towards the use of the hyphen tend to be associated with viewing nonviolence as simply the absence of violence, and the specialized literature often spells "nonviolence instead of non-violence" without the hyphen because nonviolence has positive properties beyond the refusal to use violence (Baldoli and Radaelu 2019, 1166; also see Sharp 2005).

18 See Bharadwaj for more on the relationship between charismatic leaders on practices of pragmatic nonviolence.

19 Interview by Danita Catherine Burke with Sune Scheller of Greenpeace Nordic (Denmark), September 3, 2018.

20 Interview by Danita Catherine Burke with Sini Harkki of Greenpeace Nordic (Finland), September 21, 2018.

21 Interview by Danita Catherine Burke with Sune Scheller of Greenpeace Nordic (Denmark), September 3, 2018.

22 Interview by Danita Catherine Burke with Sini Harkki of Greenpeace Nordic (Finland), September 21, 2018.

References

Ackerman, P. and C. Kruegler. 1994. *Strategic Nonviolent Conflict: The Dynamics of People Power in the Twentieth Century*. Praeger Publishers.

Ahmed, Meena. 2006. *The Principles and Practice of Crisis Management: The Case of Brent Spar*. Palgrave.

Alatas, Syed. 2007. "Eurocentrism." *Blackwell Encyclopedia of Sociology* 3(2007): 1491–1494.

Allen, J. 1979. "Anti-Sealing as an Industry." *Journal of Political Economy* 87(2): 423–428.

Atack, I. 2012. *Nonviolence in Political Theory*. Edinburgh University Press.

Baldoli, R. and C.M. Radaelu. 2019. "What Has Nonviolence Got to Do with the EU?" *Journal of Common Market Studies* 57(5): 1165–1181.

Baur, D. and G. Palazzo. 2011. "The Moral Legitimacy of NGOs as Partners of Corporations." *Business Ethics Quarterly* 21(4): 579–604.

Berlie, Laurence Schwesinger. 2010. *Alliances for Sustainable Development: Business and NGO Partnerships*. Palgrave Macmillan.

Bharadwaj, L.K. 1998. "Principled Versus Pragmatic Nonviolence." *Peace Review* 10(1): 79–81.

Blaut, James M. 1999. "Environmentalism and Eurocentrism." *Geographical Review* 89(3): 391–408.

Bowen, Frances. 2014. *After Greenwashing: Symbolic Corporate Environmentalism and Society*. Cambridge University Press.

Brolsma, Marjet., de Bruin, Robin., and Lok, Matthijs. 2019. "Introduction." In *Eurocentrism in European History and Memory*, edited by Marjet Brolsma, Robin de Bruin, and Matthijs Lok. Amsterdam University Press.

Burke, Danita Catherine. 2020. "Re-Establishing Legitimacy after Stigmatization: Greenpeace in the North American North." *Polar Record* 56(e26): 1–12.

Burke, Danita Catherine. 2021a. "The Case for a Greenpeace Apology to Newfoundland and Labrador." *The Northern Review* 51(1): 173–187.

Burke, Danita Catherine. 2021b. "The Relationship between Kanngiqtugaapik/Clyde River and Greenpeace: An Interview with Mayor Jerry Natanine." *Arctic* 74(1): 106–110.

Burke, Danita Catherine. 2022. *WWF and Arctic Environmentalism: Conservation and the ENGO in the Circumpolar North*. Manchester University Press.

Burke, Danita Catherine. 2023. *Cultural Violence, Stigma and the Legacy of the Anti-Sealing Movement*. Routledge.

Burke, Danita Catherine. 2024. "Nonviolence and Greenpeace: Apologies and Anti-Sealing Campaigning." *Arctic* 77(3): 246–260.

Carter, Neil. 2007. *The Politics of the Environment: Ideas, Activism, Policy*. 2nd Edition. Cambridge University Press.

Chenoweth, E. and K.G. Cunningham. 2013. "Understanding Nonviolent Resistance: An Introduction." *Journal of Peace Research* 50(3): 271–276.

Cianchi, John. 2015. *Radical Environmentalism: Nature, Identity and More-than-Human Agency*. Palgrave Macmillan.

Dauvergne, P. and K.J. Neville 2011. "Mindbombs of Right and Wrong: Cycles of Contention in the Activist Campaign to Stop Canada's Seal Hunt." *Environmental Politics* 20(2): 192–209.

Doyle, Timothy. 2018. "Environmental Non-Governmental Organisations." In *Handbook of Research on NGOs*, edited by Aynsley Kellow and Hannah Murphy-Gregory. Edward Elgar Publishing, 258–279.

Erickson, B. 2020. "Anthropocene Futures: Linking Colonialism and Environmentalism in an Age of Crisis." *EPD: Society and Space* 38(1): 111–128.

Essemlali, L. and P. Watson. 2013. *Captain Paul Watson: Interview with a Pirate*. Firefly Books Ltd.

Fogarty, Edward A. 2011. "Nothing Succeeds Like Access? NGO Strategies towards Multilateral Institutions." *Journal of Civil Society* 7(2): 207–227.

Fouhy, Ken. 1995. "Greenpeace Admits it Erred in Brent Spar Analysis." *Chemical Engineering* 102(10): 51.

Greenpeace Canada. 2022. "Protesting Peacefully, to Change the World." https://www.greenpeace.org/canada/en/about-us/protesting-peacefully-to-change-the-world/

Greenpeace Canada. n.d. "Our Story." https://www.greenpeace.org/canada/en/about-us/history-successes/

Greenpeace International. 2007. "Amchitka: The Founding Voyage." May 14. https://www.greenpeace.org/international/story/46686/amchitka-the-founding-voyage/

Greenpeace International. 2020. "Our Values." https://www.greenpeace.org/international/explore/about/values/

Greenpeace International. n.d. "Our Founders." https://www.greenpeace.org/international/about/founders/

Greenpeace UK. n.d. "How Greenpeace Creates Change." https://www.greenpeace.org.uk/about-greenpeace/how-we-create-change/

Greenpeace USA. 2010a. "Greenpeace Statement on Patrick Moore." July 6. https://www.greenpeace.org/usa/greenpeace-statement-on-patric/

Greenpeace USA. 2010b. "'Save the Whales' 35th Anniversary." April 27. https://www.greenpeace.org/usa/save-the-whales-35th-anniversary/

Gulbrandsen, Lars H. and Steinar Andresen. 2004. "NGO Influence in the Implementation of the Kyoto Protocol: Compliance, Flexibility Mechanisms, and Sinks." *Global Environmental Politics* 4(4): 54–75.

Harter, J. 2004. "Environmental Justice for Whom? Class, New Social Movements, and the Environment: A Case Study of Greenpeace Canada, 1971–2000." *Labour/Le Travail* 54(1): 83–119.

Hopper, Tristin. 2014. "Save (Most of) the Whales: Greenpeace Now Supports Inuit Hunting, but Native Groups Still Wary." *National Post*, August 15. https://nationalpost.com/news/canada/save-most-of-the-whales-greenpeace-now-supports-inuit-hunting-but-native-groups-still-wary

Hunter, Robert. 2004. *The Greenpeace to Amchitka: An Environmental Odyssey*. Arsenal Pulp Press.

Kashwan, Prakash and Aseem Hasnain. 2024. *Decolonizing Environmentalism: Alternative Visions and Practices of Environmental Action*. Bloomsbury.

Kerr, Joanna. 2014. "Greenpeace Apology to Inuit for Impacts of Seal Campaign." *Greenpeace Canada*, June 24. https://www.greenpeace.org/canada/en/story/5473/greenpeace-apology-to-inuit-for-impacts-of-seal-campaign/

Keziere, Robert. 2023. "Greenpeace." *The Capilano Review* 3(46): 98–103.
Knott, David. 1995. "Greenpeace Task Force Takes Brent Spar." *Oil & Gas Journal* 93(19): 25.
Kolopenuk, Jessica. 2023. "The Pretendian Problem." *Canadian Journal of Political Science* 56(2023): 468–473.
Liu, G., T. Eng and Y.K. Sekhon. 2013. "Managing Branding and Legitimacy: A Study of Charity Retail Sector." *Nonprofit and Voluntary Sector Quarterly* 43(4): 629–651.
Marberg, A., H. Kranenburg and H. Korzilius. 2016. "NGOs in the News: The Road to Taken-for-Grantedness." *Voluntas* 27(1): 2734–2763.
Mauch, Christof, Nathan Stoltzfus and Douglas R. Weiner. 2006. "Introduction." In *Shades of Green: Environmental Activism Around the Globe*, edited by Frank Zelko, Mahesh Ranagarjan, Sandra Lynn Chaney, Jane Carruthers, Peter Ho, Daniel J. Klooster, and J. Christopher Brown. Rowman & Littlefield Publishers.
Miller, Toby. 2017. *Greenwashing Culture*. Taylor & Francis.
Moore, Patrick. 2008. "Untitled Piece." *The Wall Street Journal*, April 22, A23. Republished on The Sensible Environmentalist. http://ecosense.me/2017/01/17/issues-68/
Moore, Patrick Albert. 2010. *Confessions of a Greenpeace Dropout: The Making of a Sensible Environmentalist*. Beatty Street Publication.
Morse, B.W. and M. Zakrison. 2010. "The Impact on the Inuit of Environmental Degradation to the Canadian Arctic." *Common Law World Review* 39(1): 48–68.
Paradis, Danielle. 2023. "American Anti-Pollution ad 'Crying Indian' to be Retired." *APTN News*, February 28. https://www.aptnnews.ca/national-news/crying-indian-ad-retired-us/
Pearce, Fred. 1996."Greenpeace: Storm-Tossed on the High Seas." In *Green Globe Yearbook of International Co-operation on Environment and Development*, edited by Helge Ole Bergesen and Georg Parmann, produced by the Fridtjof Nansen Institute. Oxford University Press.
Randle, M.J. 1994. *Civil Resistance*. Fontana.
Santana, Déborah Berman. 1996. *Kicking off the Bootstraps: Environmental, Development, and Community Power in Puerto Rico*. University of Arizona Press.
Shapovalova, Faria. 2010. "Arctic Petroleum and the 2°C Goal: A Case for Accountability for Fossil-Fuel Supply." *Climate Law* 10(3–4): 282–307.
Sharp, G. 2005. "Waging Nonviolent Struggle." *Porter Sargent*. https://ciaotest.cc.columbia.edu/book/ciao/0010973/f_0010973_17324.pdf
Stewart, Ben. 2015. *Don't Trust, Don't Fear, Don't Beg: 100 Days as a Prisoner of Putin – The Story of the Arctic 30*. Guardian Books and Faber & Faber Ltd.
United Nations. n.d. "Greenwashing – The Deceptive Tactics Behind Environmental Claims." https://www.un.org/en/climatechange/science/climate-issues/greenwashing#:~:text=Being%20purposely%20vague%20or%20non,and%20can%20be%20easily%20misinterpreted
Walton, Oliver Edward, Thomas Davies, Erla Thrandardottir and Vincent C. Keating. 2016. "Understanding Contemporary Challenges to INGO

Legitimacy: Integrating Top-Down and Bottom-Up Perspectives." *Voluntas* 27(2016): 2764–2786.
Weyler, R. 2004. *Greenpeace: How a Group of Ecologists, Journalists, and Visionaries Changed the World*. Raincoast Books.
Weyler, Rex. 2021. "The Roots of Greenpeace." *Greenpeace International*, June 13. https://www.greenpeace.org/international/story/48361/greenpeace-history-anniversary-founders-1971/
Wintle, Michael. 2021. *Eurocentrism: History, Identity, White Man's Burden*. Routledge
Zelko, Frank. 2013. *Make It a Green Peace! The Rise of Countercultural Environmentalism*. Oxford University Press.

3 Greenpeace and the Legacy of Its Anti-Sealing Campaigning

More than any other early campaign by Greenpeace, its work protesting seal hunting, imbedded in its work again fur, from 1976 to 1985 in Northeastern Canada (Woods 1986; Allen 1979; Burke 2021) has had the most lasting impact on its reputation in the Circumpolar North.[1] Though seconded by the legacy of its anti-whaling campaigning starting in 1975 (Weyler 2004; Hopper 2014), Greenpeace's anti-sealing legacy cemented the organization as the poster-child of activist disregard for rural and coastal peoples and their cultural heritage, knowledge, economies and ways of life (Burke 2020, 2021). It is worth noting, though, that Greenpeace's work against sealing was part of a wider movement which sprung to international attention galvanized by the leadership of the International Fund for Animal Welfare starting in 1967 (Davies 1970; Allen 1979; Patey 1990)[2] which Greenpeace later joined and arguably revolutionized with its image events and direct action activism which captivated the global media (Allen 1979; Wenzel 1987).

Greenpeace transformed campaigning tactics and image events with its "[s]kilful manipulation of the mass media", which helped it "engender enormous sympathy for their cause" (Kalland 2009, 82), putting both the organization and its messaging front and centre in the global campaign against seal hunting. Greenpeace was instrumental in the framing and normalization of anti-sealer rhetoric, actions, attitudes, and imagery (Allen 1979; Patey 1990; Harter 2004; Kalland 2009); something that continues to cause cultural and economic violence for the targets of these campaigns decades after Greenpeace formally withdrew from activism on the subject (Burke 2021, 2023, 2024). As this chapter illustrates, the complexities of Greenpeace's anti-sealing legacy is compounded by a number of factors, such as changing agendas involving stakeholders and rightsholders in the Circumpolar North; the internal politics within Greenpeace which impacted how its seal campaign unfolded; the popularity of its anti-sealing work with Greenpeace's support base in contrast with its unpopularity amongst many northern-based audiences; and how Greenpeace's attitudes, actions

DOI: 10.4324/9781003674023-3

and behaviours at the time of its campaigning reflected its relationship with nonviolence philosophy.

First and foremost, Greenpeace's anti-sealing legacy is a cautionary tale in how preoccupation with infighting, lack of accountability, and commitment to a grounded identity and set of principles can create cascading issues for an organization decades after it has resolved its internal issues and stopped actively campaigning on a topic. It is through recognizing this cautionary tale for Greenpeace's legacy that its subsequent achievements in the Circumpolar North can be understood, in part, as they reflect some lessons learned from the destructive past of Greenpeace's anti-sealing work.

The Rise of the Anti-Sealing Movement

Traditionally, seal hunting has been a common practice by many Indigenous and non-Indigenous cultures and communities for centuries, with seal hunting in the Cirumpolar North and North Atlantic being the most commonly known. The anti-sealing movement started in Canada (Allen 1979; Burke 2022b),[3] but the campaigning impacted fishers, their families, communities, and cultures across the Circumpolar North, including places like Kalaallit Nunaat/Greenland, Finland, Sweden, and Norway (European Commission 2023; also see Allen 1979; Burke 2020; Svels et al. 2025). Rural and coastal communities in the Canadian North and Canadian Northeast practised seal hunting for generations, with most commercial and traditional seal hunters residing in Newfoundland and Labrador, the Quebec North Shore, the Magdalen Islands, Nunavut, and the Northwest Territories (Woods 1986; Sinclair et al. 1989; Hawkins and Silver 2017; Burke 2021).[4]

Movement against commercial seal hunting started to get traction in the 1950s as activists and media began to highlight the killing of seals and criticize the methods of hunting and the use of animals in fashion (Wenzel 1987). The awareness of and push against seal hunting, however, exploded in 1964. In 1964, a French Canadian documentary falsely claimed that seals are skinned alive and that this is a standard practice in seal hunting. The documentary showed a person skinning a seal alive. The purpose of this demonstration was to convey what the filmmakers wanted to be seen as the typical – though false – way to skin a seal during the seal hunt (CBC Radio 1965; Îles de la Madeleine 2021). It later emerged that the individual who skinned the seal for the filmmakers was expressly paid to do it in this misleading manner by the documentary crew and that it occurred outside of the seal hunting season (Burke 2021). The outrageous and inhumane imagery of the seal being skinned in this fashion was an extremely powerful image. It was instrumental in galvanizing people against sealing despite it being a false depiction of standard, accepted sealing practice.

What resulted was the rise of the anti-sealing movement as we know it today, starting in 1967 with the "Save the Seals" campaign (Allen 1979). This developed into the International Fund for Animal Welfare (IFAW) in 1969, which was founded explicitly to protest the seal hunt (Davies 1970). The then new Canadian province of Newfoundland [and Labrador], which had joined Canada in 1949, was particularly highlighted and targeted by campaigners, many of whom initially came from mainland Canada (Patey 1990; Burke 2023). Quebec's Îles de la Madeleine (Magdalen Islands) in the Gulf of St. Lawrence was also targeted extensively by activists (Îles de la Madeleine 2021; Burke 2023).

Though the anti-sealing movement superficially appears as a monolith, in fact, it was a broad collection of actors all converging against aspects of seal hunting around the same time (Phelps Bondaroff and Burke 2014).

> Initially, the anti-sealing movement had two primary goals. While some organizations, including the Audubon Society, the WWF, and the Canadian Wildlife Federation, focused on the harp seal's long-term survival, others, especially animal welfare organizations such as the RSPCA and HSC, placed emphasis on the way the hunt was conducted. Both groups of organizations ran information campaigns, displaying graphic images in order to raise awareness of the issue through the media. This was accompanied by a letter-writing campaign which took advantage of the large memberships of these organizations.
>
> (Phelps Bondaroff and Burke 2014, 170)

While there is no agreed-upon timeframe for the anti-sealing movement, it is broadly viewed to be at its most prolific between 1969 and 1983, when the European Economic Community established its first ban on white coat harp seal products (European Commission 2023).[5] Anti-sealing activism continues to the present time. Organizations like the IFAW, Sea Shepherd Conservation Corps, and the People for the Ethical Treatment of Animals continue to speak out against seal hunting, though with less on-the-ground activism in recent decades. The large-scale movement of various activist organizations and their supporters against sealing, which dominated in international media in the 1970s–1980s, greatly subsided after the implementation of the second European ban on all commercial seal products in 2009 (European Commission 2023).

Greenpeace's Anti-Sealing Activism: A Brief Overview

The issue of seal hunting became one of Greenpeace's first large post-Amchitka campaigns, starting in 1976 (Allen 1979; Phelps Bondaroff and Burke 2014). Initially, Greenpeace pursued a sustainable use-oriented agenda, which included proactively seeking out a relationship with sealers in Newfoundland and Labrador, as both parties wanted to curtail commercially focused foreign overhunting (Harter 2004). At first, Greenpeace recognized the culturally distinct and close relationships between Newfoundland and Labrador's people and Inuit with seal hunting and focused on the issue of foreign overhunting of the harp seal herd (Harter 2004; Burke 2020).

Shortly after getting involved, Greenpeace convinced the Newfoundland Fishermen, Food and Allied Workers Union, which represented Newfoundland fishers in 1976, to cooperate with them against Norwegian sealing vessels that were hunting commercially off the Canadian coast (Harter 2004). Greenpeace acknowledged Inuit and Newfoundland and Labrador sealers' positive relationship with seals and the importance of seals to local peoples; "The fact is that the commercial fleets owned by Norwegian companies are wiping out the seal herds…while the fishermen and [Inuit] of Newfoundland and Labrador do eat the meat" (Greenpeace Chronicles 1976, 6 as quoted in Harter 2004, 96). However, the alliance fell apart quickly. By 1977, Greenpeace shifted "its focus from an endangered species argument to outright opposition to all sealing" (Wenzel 1987, 199).

According to Allen (1979, 426), who wrote contemporaneously at the height of the anti-sealing movement, Greenpeace and IFAW were innovative in their activism. The shift away from local ally-ship coincided with the profitability – financially and in terms of international reputation and profile – of anti-sealing protests for Greenpeace. Allen (1979, 426) states that Brian Davies, founder of IFAW, discovered that

> While the seal hunt had held little interest to anyone outside of Newfoundland, his appearance at the hunt, in protest, was considered sufficiently news-worthy to warrant extensive coverage in the news media. By simply appearing, Davies provided his firm with enormous quantities of information, distributed at almost no cost to the firm. The introduction of this innovation assured the firm of quick success… Greenpeace Foundation cleverly modified Davies's innovation in a way which allowed them to break down this particular entry barrier to the anti sealing industry. They not only sent representatives of the firm to the hunt, but, by promising to perform outrageous acts which would guarantee an interesting story.

Allen observed at the time that Greenpeace's involvement in anti-sealing protests helped it to significantly grow as an organization on the national and global stage in the three-year period from 1977 and the publication of Allen's article in 1979.

The quick shift in Greenpeace's position on seal hunting and seal hunters coincided with its internal power struggle, leading to the organization targeting its former allies (Patey 1990; Dale 1996).[6] This dynamic occurs largely as a result of the increasing number of animal rights activists becoming involved in Greenpeace and the rise of Paul Watson as their leader.

Frank Zelko (2013, 8), in his work on the rise of countercultural environmentalism, states that

> The antisealing [sic] protests attracted a substantial number of animal rights activists to Greenpeace. From an animal rights perspective, there can be no question of compromise when it comes to the killing of whales or seals – abolition is the only goal worth pursuing.

An alliance with local sealers, therefore, was untenable for individuals within Greenpeace who viewed any killing of seals as unacceptable; a segment of the organization with a growing influence in the 1976–1977 period during the rising influence of Paul Watson within the organization.

At the same time as animal rights activists began to dominate Greenpeace internally, Greenpeace became

> The first group to employ direct action on the ice, with some of its activists physically intervening to prevent sealers from bludgeoning seals by lying across the seals or by removing seals from the reach of sealers…. [and] blocking and to chaining themselves to sealing vessels.
>
> (Phelps Bondaroff and Burke 2014, 171)

Jessica Wilson of Greenpeace International, and formerly of Greenpeace Canada,[7] recalled that

> Paul Watson…and Bob Hunter at the time were the de facto leaders of Greenpeace and they started this seal campaign…Bob Hunter and Paul Watson end up having a difference of opinion…Bob Hunter wanted to go what he called deep green, less about animal rights and more about the notion of pure ecology, which involves humans.[8]

However, Watson pushed the organization further down an animal rights-focused path. Zelko (2013) notes that Watson and Walrus Oakenbough, a fellow Greenpeace member, pushed the organization to use increasingly confrontational tactics when dealing with local fishers and to be more hard-lined against seal hunting (Zelko 2013, 236).

Watson and his supporters used what could be described as monkey-wrenching-style tactics.[9] These tactics included destroying the equipment of sealers, such as hakapiks and gaffs, which are both traditionally part of hunting equipment and safety gear to navigate unstable and slippery ice sheets as part of a strategy of intimidation (Patey 1990). Rapidly, the organization found itself "pitted against rural working-class or indigenous people struggling to make a living" (Zelko 2013, 232), inverting the David versus Goliath narrative.[10]

For example, Greenpeace members reportedly abused sealers; now it was the Goliath attacking vulnerable working-class fishers under the influence of the animal rights activists. There is one reported instance where four Greenpeace activists out on moving ice kept a sealer hostage during the hunt, at some point during the 16th–19th March 1977 time period. According to Harry Roswell, a veterinary pathologist who was an observer of the seal hunt in 1977 and 1978 for the Committee on Seals and Sealing in Canada (Roswell 1977, 1978),

> Four of them [Greenpeace members] surrounded a sealer and refused to let him move from the pan. This intimidation can only be considered as bullying. In spite of the intimidation, the harassment and the provocation, the Newfoundland sealer did not strike out against Greenpeace members or attempt to take any form of defensive action. This type of conduct must receive the commendation of the people of Canada as one of the few of them, placed in a similar circumstance, [who] would exercise [such restraint].
>
> (Roswell 1977, 25)

Roswell went on to report that the activities of Greenpeace in 1977 on the ice were conducted under Paul Watson's direction, which created situations that "were nothing short of bullying, intimidation, hostage taking and physical and psychology harassment" (Burke 2021, 178).

For his part, Watson also reportedly "faked being paralyzed after a direct action stunt he instigated caused an accident" (Burke 2021, 178). During the incident, Watson connected himself to a load of pelts being hoisted up onto a vessel via a wench and was mistakenly dropped during the incident. Watson's drop caused alarm and distress to the wench operator, who feared that they would kill or permanently maim him; a situation which Watson exploited. Using the concern of the vessel's crew, Watson claimed to be paralysed to gain access to the vessel for medical help, and once on board, he began to walk around and taunt the sealers with threats of legal action (Roswell 1977, 23). Watson, however, recounts his anti-sealing campaign experience while working as part of Greenpeace as evidence of historic nonviolent intervention (Essemlali and Watson 2013; also see Weyler 2004).

Ultimately, labelling Greenpeace's tactics during the 1977 hunt as nonviolent direct action is very questionable; even leadership at the time recognized that (Burke 2024). Sharp (2005, 8) states that nonviolence is "identified by what people do, not by what they believe". In some ways, Greenpeace's work against seal hunting, and to what extent their conduct may be considered violent, is open to interpretation. It is clear, however, from actions taken by Greenpeace leadership at the time that even the organization's own leaders knew that what had happened during the 1977 hunt went too far.

Watson's leadership and activities in 1977 were the final straw for his time with Greenpeace. As a result of his activities, he was removed from the organization. As previously stated, Weyler observed that Watson "tended to push the end of non-violence", and he was removed from Greenpeace as the leadership recognized he was undermining Greenpeace's nonviolent reputation with his activities against sealers (Weyler 2004, 457).

Reflecting on Watson's legacy and its implications for Greenpeace, one Greenpeace member stated,

> The leaving of Paul Watson doesn't change the fact that we did have an anti-sealing campaign in the late 1970s and early 1980s in the Arctic. I think that campaign was, and we've also tried to explain on several occasions, handled [in such a way that] mistakes were made....But the consequences of the campaign, no matter the intentions, had the consequence of a huge decline of the Indigenous communities being able to sell off, for example, the fur from sealing and that was due to our poor handling and unintended consequences. Nonetheless, still it was poorly communicated. Of course other actors joined that campaign and made it harder for us to control [the messaging] in that way but I wouldn't put it, it wouldn't absolve us but I wouldn't blame it as such on Paul Watson because it was a decision made by the organisation to run that campaign. Maybe he was the most vocal. I have no idea what might of went down back then [but]...it keeps coming up and it's been a really really long progress to try to, I don't know if I want to say make amends but at least explain an apology for the mistakes that were made, and we did.[11]

Despite recognition from at least one Greenpeace member that Watson's leadership, and subsequent departure from Greenpeace, does not absolve the organization for questionable elements of its campaign, the desire to make amends for that legacy is hesitant. As of the publication of this book, Greenpeace has not apologized to the people targeted by Watson and the Greenpeace members he led during the 1977 direct action on the ice.

The outcomes from the direct action brought notoriety to Greenpeace (Allen 1979; Patey 1990; Harter 2004), but it has also brought condemnation (Patey 1990; Burke 2022b, 2023) and is part of the legacy of colonial attitudes, actions, and behaviours for which Greenpeace is trying to make some amends (Kerr 2014). Jim Winter, a former CBC journalist and founding president of the Canadian Sealers Association, reflected that

> I would say that the legacy is the selective use of language and images to convince urbanites, and even some rural peoples, that sealers are barbarians; that we lacked their understanding of civilization. Seals were portrayed as different from other animals to convey that only ignorant people could kill them. All the "factoids" peddled about sealing caused people to condemn sealers and our families and communities as subhuman.
>
> (Burke 2022a)

The direct action against rural sealers in 1977, in particular, illustrates that Greenpeace struggled to remain committed at times to a philosophy of principled nonviolence, likely as the organization internally was struggling to pinpoint its direction. Its failure to apologize at the time, and to this day, for these incidents also calls into question the organization's commitment to taking responsibility for its actions.

Instead of addressing its failure to live up to principled nonviolence (Bharadwaj 1998), the organization was more focused on keeping itself together and navigating financial issues. By 1978, the organization was reportedly experiencing financial difficulties after years of steady campaigning. At the same time, the anti-sealing protests were purportedly good fundraising opportunities, especially for Greenpeace's California-based members who were against all seal hunting (Zelko 2013).

In a very frank interview in 1978 with the CBC Paul Watson, shortly after being expelled from Greenpeace, was very open about the financial motivations behind Greenpeace's early seal campaigning and the disconnect between supporters and the realities of seal hunting:

> A dozen people this year from Greenpeace California, I mean they're coming from the highest standard of living region in North America, they're travelling to the place with the lowest income per year on this continent telling them not to kill seals because they're cute. But they're not an endangered species.
>
> (CBC 1978)

Watson also stated that

> I think that of all the animals in the world or any environmental problem in the world the harp seal is the easiest issue to raise funds on. Greenpeace has always managed to raise more money on the seal issue for campaigns than has actually been spent on the campaigns themselves. The seal hunt has always turned a profit for the Greenpeace Foundation.
>
> (CBC 1978)

CBC journalist Barbara Frum, who interviewed Watson, inquired if the seal activism was connected to Greenpeace's need for funding. Watson replied: "definitely because it's easier to make money and because it does make a profit. There are over a thousand animals on the endangered species list, and the harp seal isn't one of them" also noting that "the seal is very easy to exploit as an image" (CBC 1978; also see Zelko 2013 for more on the Greenpeace's in the late 1970s and the financial pressures it was under). Ultimately, there is strong evidence to suggest that Greenpeace's internal battles for dominance and financial difficulties impacted its commitment to nonviolence and the handling of its involvement in the seal protests.

The 1977–1979 period was the height of Greenpeace's profile for its anti-sealing activism (Allen 1979; Patey 1990; Harter 2004; Zelko 2013), but by 1986 it formally stopped participating in anti-sealing campaigning. It stopped campaigning against seal hunting reportedly due to coming to realize the negative impact of the collapse of the commercial seal hunt on Indigenous peoples (Woods 1986).

The Inuit of Canada, for example, most prominently in the Northwest Territories and Nunavut, were severely impacted by the decimation of the sealing industry in the 1980s (Woods 1986), starting with the fallout from the first European ban on seal product imports in 1983 (European Commission n.d.).[12] As a result of the first European seal product ban, Wenzel (1987, 197) acknowledged that

> In this changed economic environment, the ringed seal not only held its original subsistence value, but within a changed demographic and technological situation, such as developed in Arctic Canada from the 1950s onward, it also became the main cash-producing commodity available locally to Inuit, if only because of its temporal and quantitative availability…The economic condition of Inuit sealing, beyond a basic cash value per sealskin, has, however, only recently come to be better appreciated. The inherent instability of reliance on a single marketable species…with little or no ability for the producers to influence that market…had a serious effect on the maintenance of the Inuit seal hunting.

In sum, Inuit were disproportionately impacted compared to many non-Indigenous coastal peoples in Canada because the sale of seal products made up a large percentage of the economy in Inuit communities where economic opportunities were more limited (Woods 1986; Phelps Bondaroff and Burke 2014). While Greenpeace came to realize the severity of the negative impact on Indigenous peoples, namely, Inuit, it was inconsistent in staying away from its anti-sealing past, and it was not until 2014 that it articulated a concise, though limited, apology to Inuit, Indigenous, and coastal peoples in Canada for elements of its seal protest past (Kerr 2014).

Indigenous Leadership Pushing Greenpeace's Change

Greenpeace formally withdrew from anti-sealing and anti-fur trapping activism more broadly by 1986 (Woods 1986), which appears to be just as the organization began its process of professionalization which saw it move away from campaigning approaches from the past (Carter 2007). This marked the beginning of the organization's long and ongoing road of reconciling with its anti-sealing legacy. But the move to stop protesting happened at the urging of Indigenous peoples, predominantly Inuit based in Canada, Kalaallit Nunaat/Greenland, and Alaska, as well as Indigenous peoples from across Canada and the decision to stop this campaigning was not an easy one for the organization to reach.

By the mid-1980s, a sense of urgency had arisen amongst Indigenous leaders, communities, and researchers, and civil servants aware of the on-the-ground situations to protect Indigenous peoples' economies and societies across the Circumpolar North from the mounting adverse impacts of anti-fur protests, especially in Inuit communities based in Alaska, Kalaallit Nunaat/Greenland, and Canada. Terry Fenge, a prominent actor in Arctic affairs around this time, wrote to the editor of *The Guardian* newspaper in the United Kingdom in an effort to try to get the media to understand the severity of the situation that they helped foster:

> Should Greenpeace UK and other like-minded groups succeed with their anti-fur campaign, the umbilical cord between "the land" and native peoples will be cut and, in time "the land" will become for them what it is for most of us – a factor of production. Who then in the vast expense of northern Canada will, to quote the Canadian National anthem, "stand on guard" for the environment?
>
> (Fenge 1985, 2)

The organization, Indigenous Survival International (ISI),[13] formed in this period and was prominent in the mid-1980s, calling out Greenpeace and other anti-sealing and anti-fur protesting organizations for their severe impact on Indigenous peoples and the hypocrisy of the organizations and their supporters.

> The survival of the Indigenous People of Canada, Alaska and Greenland is threatened by the neo-colonial animal rights movement. We call it colonialise because: it is destroying our culture, economies and social institutions; it is violating our fundamental human right to self-determination and economic self-sufficiency; it is based ethnocentric assumptions; and it is exercised by a force which is foreign to us. We call it colonialism because we are familiar with this concept: the Indigenous People ... have been victims of European territorial colonialism for over

> three centuries. It is ironic, indeed, that that which was so instrumental in this first wave of colonialism – the fur trade – is precisely what our colonizers and would-be colonizers are now trying to squelch.
>
> (Indigenous Survival International 1985, 1)

The anger and upset at the reiteration of colonialism in the form of animal rights groups further undermining Indigenous peoples across the North was profound, as Indigenous, as well as non-Indigenous, suffered a lot of harm as a result of these protests, though it largely happened out of the gaze of the media, and therefore went unreported (CBC The Broadcast 2021; Woods 1986; Allen 1979).

Inuit communities, in particular, were severely impacted by the fallout from anti-sealing and anti-fur trapping and trading activism. The Fur Institute in Canada, for example, reported in 1988 that

> The loss of trapping would have extremely grave consequences for aboriginal and other cultures in Canada. The devastating impacts of the European Economic Community ban on whitecoat seal pelts on high Arctic Inuit communities represent only a taste of the social problems, such as increased incidence of alcoholism and suicide, that would result from the loss of the fur trade.
>
> (Fur Institute of Canada 1988)

Inuit and Indigenous leaders and their supporters worked to minimize the continuing fallout as observed and reported by the Fur Institute, as they had direct awareness and experience of the socio-economic and psychological harms facing their communities.

Given Greenpeace's high profile and leadership in the anti-sealing and associated anti-fur campaigning, Greenpeace was targeted for dialogue with the hope to encourage the organization to acknowledge the harms happening on the ground as a result of activism they helped normalize. Local activists hoped to get Greenpeace to reconnect with their stated commitment to nonviolence and respect for Indigenous peoples and cultures by helping to find solutions to the issues that had arisen. ISI had a central role in confronting Greenpeace's legacy in anti-sealing/anti-fur trading and it observed in the lead-up to getting Greenpeace to modify its position in the 1980s:

> In our view, the only thing immoral about fur is the devastating impact that a campaign against its trade will have on the Indigenous Peoples of this world. Once again, as in the anti-seal campaign, it will be the most innocent and most susceptible who will be the hardest hit.
>
> (Indigenous Survival International 1985, 3–4)

But as documents from the Laurier Archives based at Wilfrid Laurier University in Waterloo, Canada reveal, internal tension within Greenpeace, specifically resistance from its office and members based in the United Kingdom, was a big barrier to the organization stopping active anti-sealing and anti-fur campaigning. At the same time as Greenpeace was negotiating with ISI, which was supported by the World Wildlife Fund (WWF),[14] to formally withdraw from active anti-sealing and anti-fur trading work, its UK office was full-steam ahead on that very agenda, undercutting bilateral talks.

As ISI (1985) reflected,

> In Canada…we have higher rates of suicides, alcoholism and violent deaths…Some communities experience unemployment levels as high as 90%…Many other circumpolar Indigenous communities in Greenland and Alaska are suffering in the same way. We understand that the animal rights activists are now turning to furs and trapping. The RSPCA, here, in Britain, have apparently launched its crusade focussing on the "morality of wearing furs". And Greenpeace (U.K.) is campaigning for a "fur-free" Britain.

Interestingly, and largely unacknowledged about this time period in present-day retelling of Greenpeace's anti-sealing legacy is that within the broader organization, Greenpeace's Danish office,[15] Greenpeace (Denmark), appears to be early supporter for Inuit subsistence hunting; internally, the office was a leader on the subject of acknowledging and protecting Inuit traditional rights on the sealing issue.[16] Greenpeace (Denmark) leadership in paving the way for Greenpeace to start being more nuanced in its approach to sealing was acknowledged by ISI as it worked to negotiate with Greenpeace about its stance within Canada, and globally:

> Greenpeace (Denmark) has endorsed the traditional seal harvest by the Inuit of Greenland. In Canada, we have the support of the World Wildlife Fund (WWF). Very recently, in Canada, a meeting was convened between ISI, Greenpeace and WWF to begin discussions on how we might sort out our differences on the animal rights matter.
>
> (Indigenous Survival International 1985, 11)

But despite movement within Greenpeace, particularly coming from Denmark, to evolve the organization's stance on sealing, Indigenous leaders in Canada reflect that, in their experience in 1985 at least, Greenpeace was not open to their voices. In a press release in response to Greenpeace's engagement with Indigenous leaders in Canada through ISI in 1985, leading up to its eventual withdrawal from the anti-sealing movement in 1986, it was noted:

> It appears that the Green Peace [sic] movement, which makes great mileage out of Aboriginal people and Aboriginal culture, is not as receptive to meeting with people at the grass roots level as the public pronouncement would seem to indicate.
>
> (Laurier Archives CARC holdings 1985, 2)

Greenpeace's resistance to Indigenous concerns in 1985 about its anti-fur trading stance while continuing to make public associations between Indigenous imagery and culture with its organization concerned key Indigenous leaders in Canada working on dialogue with Greenpeace and protecting Indigenous hunting rights. Among those concerned included Georges Erasmus, national chief of the Assembly of First Nations and co-chairman of ISI, as well as Sam Sinclair and Bob Stevenson.

This concern is illustrated, in their own words, in the ISI Canadian Sub-Committee meeting notes from 1985 (ISI Canadian Sub-Committee Meeting 1985):

Bob Stevenson: "ATFC [Aboriginal Trappers Federation of Canada] incorporated last June and new members are joining every month. ATFC was upset about Greenpeace – is it more important than ATFC?"

Georges Erasmus: "That did not enter into it. The meeting with Greenpeace fulfilled an obligation of founding conference. Suggested to Greenpeace that they meet with ATFC and Inuit. Greenpeace was very nervous about meeting at all, saw ATFC as the group most affected and thus were more comfortable with ISI, an umbrella group. Greenpeace will meet Inuit next week in Rankin Inlet, and are ready to meet ATFC. Press statements: why say we are fighting? ISI has no dispute with ATFC. Nobody gains. It remains in the public image that native people aren't even together on a non-political issue like saving their cultures"

Bob Stevenson: "Wants ISI to let groups like Greenpeace know that ATFC must be dealt with, too. Going to press lets ATFC membership know ATFC is not in agreement with things like Greenpeace meeting".

Georges Erasmus: "Probably Greenpeace are playing games and are afraid of meeting with trappers. Whatever is in the future, let's work it out internally".

Recognising the importance of swaying Greenpeace to their cause, Indigenous leaders in Canada, through ISI, persisted in reaching out, but as the Laurier Archives files reveal, Greenpeace both wanted to be seen to engage Indigenous leaders, but were resistant at the same time to what Indigenous leaders were trying to get them to understand and take onboard.

In 1986, ISI's work to engage Greenpeace progressed with a meeting in the autumn of that year "to be informal for the start of our dialogue" (Gilday 1986, 3). ISI, however, observed Greenpeace's reluctance to engage with them in the way that they had conveyed prior in preparation for the meeting, as noted by Cindy Gilday of ISI in a letter to Benjamin P. Nageak of the Department of Wildlife Management, North Slope Borough, Alaska. At least from ISI's perspective, there was a sense of mixed messages from Greenpeace leadership:

> Greenpeace International Board has appointed Bertil Hagerhall to meet with ISI. I was told he is the chairman for Greenpeace International Committee on Natural Resources. Obviously the message given by Patrick Moore from Vancouver and Ann Dingwall from Washington, that we would be meeting the full committee from different countries has been changed.
>
> (Gilday 1986, 2)

Though now disavowed by Greenpeace, at the time of dialogue with ISI, Patrick Moore was a key player both in Greenpeace and in making the negotiation process with ISI happen.

Moore was once a prominent Greenpeace member in the 1970s and early 1980s (Weyler 2004; Zelko 2013), including in its anti-sealing activism (Roswell 1977, 1978). Monte Hummel of WWF, who supported ISI, issued a statement to the Aboriginal Peoples Conference on Harvesting and Renewable Resources on January 30 to February 1, 1986, acknowledged Greenpeace's reluctance in engaging with Indigenous peoples about sealing and fur and highlighed Moore's importance in bridging this communication gap. Hummel states, in his view, that

> I know that Greenpeace is not eager to be in conflict with you. They would much rather work on issues they feel you both have in common, such as militarization of the North. I also think that Greenpeace's withdrawal of the "fur-free Britain" campaign was a very important gesture on their part. It caused considerable debate within Greenpeace. It resulted in losing some supporters in the U.K., and it developed some distance between Greenpeace and the most extreme European anti-harvesting groups. I believe Patrick Moore was crucial in persuading Greenpeace to make this move.
>
> (Hummel 1986, 3)

Hummel also made a point to stress that: "In my view, these groups are at their worst when demonstrating their profound intolerance of any views other than their own. Unfortunately this often translates into intolerance for cultures other than their own".

The theme of activist intolerance about cultures and peoples different from their own is a recurring part of Greenpeace's relationship with its anti-sealing legacy, as the next section further demonstrates. At this point, a key takeaway is that there was awareness in the 1980s that Greenpeace and other protest organizations and their supporters were demonstrating intolerance, especially amongst Indigenous communities and leadership in Alaska, Kalaallit Nunaat/Greenland, and Canada. Furthermore, Greenpeace, despite its contemporary emphasis on support for Indigenous traditional hunting, did not give up its anti-sealing and anti-fur activism without resistance; it was pressure from social changes and the rising voices and authority of Indigenous peoples, especially Inuit, which appears to have brought about this shift.

Reconciling with Greenpeace's Anti-Sealing Past

Over 35 years after Greenpeace started protesting against seal hunting, members of the organization in offices in Canada and the Nordic region recognized that more should be done to address the organization's problematic conduct during its anti-sealing campaigning, focusing on the impact on Inuit traditional hunting rights (Kerr 2014; Burke 2020).[17] Reflecting on Greenpeace's past anti-sealing activism, Mads Flarup Christensen of Greenpeace Nordic also noted:

> I think we got things wrong in the 70s, for example, around the sealing issue where I think we did not have sufficient knowledge or grounding in those areas and communities to really be able to, as city people and as people from other countries, to go into areas and have a massive impact on local life, without being grounded and founded there. I think we burned our fingers on that and understood that this is not the way we could work and therefore I think from the beginning 2000s when we increasingly saw that we had another frontier coming up in the Arctic around both as a consequence of climate change but also because of the resource grab that was starting up in the region, that this would be a new frontier for us.[18]

Jessica Wilson, formerly of Greenpeace Canada, further stressed: "I don't think we can understate the depth of the hatred and resentment between Inuit and Greenpeace" largely as a result of the anti-sealing legacy.[19]

This self-reflection by Greenpeace members who have been active in their northern engagement with rightsholders and stakeholders in Kalaallit Nunaat/Greenland and Canada about the lasting impact from their anti-sealing activism is a positive step in building inroads with peoples across the Circumpolar North; to address a problem like a bad reputation, you need to recognize why it exists in the first place. This period of reflection by some within Greenpeace is also happening against a backdrop of ongoing changes in the governance structures and norms of the Canadian and Kalaallit/Greenlandic Arctic (Phillips 2010; Loukacheva 2012; Kuokkanen 2017; Rodon 2017; Hofverberg 2019). Governance and legal changes have increasingly empowered Indigenous peoples and coincide with their mounting demands for reconciliation for colonialism and demands for a shift in attitude and thinking about Indigenous peoples (Qikiqtani Inuit Association 2013; Kuokkanen 2017; Rodon 2017; Crown-Indigenous Relations and Northern Affairs Canada 2019, 2022; Van Holstein and Head 2018). Nowadays, the pursuit of legal and political changes through activist campaigns that directly impact Indigenous peoples, cultures, and territories without local Indigenous involvement is effectively a non-starter for an organization. A legacy like Greenpeace's, therefore, is a barrier with some Indigenous peoples in the Circumpolar North, and going it alone in northern and Arctic activism is an ineffective option.

Environmental and animal rights organizations are part of the fabric of colonialism, which Indigenous governments and peoples are working to confront and address as part of their processes of decolonization and reconciliation (Erickson 2020; Van Holstein and Head 2018).

> ENGOs … were initially founded primarily in Europe and Northern America beginning in the 19th century … [and] they emerged out of European colonial relations. Modern environmentalism has from the beginning mixed Northern concerns about wilderness preservation with scientific and managerial priorities of colonial government.
>
> (Partelow et al. 2020, 12)

For its part in 2014, Greenpeace Canada offered an apology to Inuit, Indigenous, and coastal peoples in Canada for its anti-sealing campaigning and the attitudes and actions based on colonial thinking that spurred the anti-sealing movement (Kerr 2014), though the apology is light on detail about what it specifically did and to whom.

Furthermore, Greenpeace's apology predominantly focused on the Inuit. Kerr (2014) states,

> Though the campaign was directed against the commercial hunting of seals — and not the small-scale, subsistence hunting carried out by Northern Indigenous and coastal peoples — we did not always communicate this clearly enough. The consequences of that, though unintended, were far-reaching.

In it, Kerr stresses on behalf of Greenpeace that it did not campaign directly against Inuit hunting but was neglectful to highlight and differentiate between different forms of seal hunting and caused unintentional harm and hardship, such as economic loss from the collapse of prices for seal pelts (Kerr 2014; also see Woods 1986; Bernauer 2020 for background).

Kerr (2014) goes on to state that Greenpeace

> Acknowledge[s] the role we played in the unforeseen consequences of these [seal product] bans. In 1985, we issued an apology for our role in this. Now we must go further. Like the corporations we campaign against, we too must be open to change. Open to examining ourselves, our history, and the impact our campaigns have had, and to constantly reassessing ourselves — not just by apologizing, but by humbly making amends and changing the way we work. And we have a responsibility — not just as an organization that once campaigned against the commercial hunt, but also as conscious, socially responsible human beings — to right wrongs, to actively stop the spread of misinformation, and to decolonize our thinking, our language and our approach.

Greenpeace's effort to reconcile with Inuit in particular is a complicated path to navigate, given the sort of things that Greenpeace would need to do to make itself appealing to Inuit audiences and the problems that such actions and changes in attitudes and thinking would present for Greenpeace's relationship with its primary support and donor base (Burke 2020).

For example, in 2010–2011, Greenpeace Nordic (Denmark) member Jon Burgwald was tasked with helping Greenpeace establish dialogue with the Inuit of Kalaallit Nunaat/Greenland, which included addressing the organization's anti-sealing legacy and its implications for local peoples head-on. As Christensen recalled,

> Our approach there was from the beginning was to partner up and do what we do in alliance with Indigenous organiations. I think we [Greenpeace] Nordic have done that… increasingly in Greenland and elsewhere from the late 2000s when we really started to move into the Arctic…Of course, it is not easy because we did not have a very good relationship with, or a good brand name, due to our mistakes and our way of operating in the late 70s and the beginning of the 80s and because some people have been very successful in sort of projecting everything that was going wrong in for example Greenland or in some Indigenous communities pointing that to be based on one symbol, namely Greenpeace.[20]

As part of this work, Burgwald publicly wore a seal skin vest to signal support for Inuit subsistence sealing while doing a media interview criticizing Sea Shepherd (Burke 2020).

Burgwald's efforts were positively received by Inuit in Kalaallit Nunaat/Greenland, but once it became news in mainland Europe, Greenpeace membership was shocked and offended by Burgwald's activities. Burgwald's activities were perceived to be a reversal of Greenpeace's anti-sealing work, a campaigning legacy that many Greenpeace supporters like a lot (Burke 2020).

Faiza Outahsen of Greenpeace Netherlands recalls: "[Jon Burgwald] was … called out here internally, but Jon also had a very difficult time and encountered a lot emotions" as local people threw rocks and eggs at him when in Nuuk, Kalaallit Nunaat/Greenland.[21] Outahsen also noted that

> The whole seal thing it did also gain attention, why is someone from Greenpeace wearing sealskin? Obviously a lot of people that support us … because of the seal campaign so this is why it is something that you have to explain in your position because it is very much about the industrial type of, and not what the Indigenous communities do in a sustainable way.[22]

Greenpeace's internal reaction to public efforts to acknowledge Inuit sealing rights is an uncomfortable position for some base supporters and offices outside of the Arctic region (Burke 2020). There is a desire to support Indigenous rights and confront elements of Greenpeace's contribution to the colonial legacy of environmentalism (Kerr 2014; also see Greenpeace USA n.d.; Greenpeace Nordic 2013; Greenpeace Canada 2017; Thoener 2024). Christensen acknowledged that, in his opinion, Greenpeace has: "quite good principles and policies in place when it comes to what, in the ways we work with these, what special rights we think Indigenous peoples have, that perhaps we don't recognize elsewhere".[23]

For example, in addition to the Greenpeace Canada apology for aspects of its anti-sealing legacy in 2014, Greenpeace has been trying to develop better relations with Indigenous peoples in the Circumpolar North, as demonstrated through the development of an Indigenous Peoples Policy. Greenpeace's Indigenous Peoples Policy was developed out of the organization's efforts to engage Arctic Indigenous peoples in 2013 in Kiruna, Sweden; a meeting with Indigenous peoples' representatives Greenpeace arranged to coincide with a major Arctic Council meeting in Kiruna.

Greenpeace Nordic's (2013) 2013 annual report states of the Kiruna meeting:

> 'On May 15th, Greenpeace gathered indigenous people from all over the Arctic, the USA, Russia, Canada, Greenland, Sweden, Norway

> and Finland for the conference 'Peoples Arctic – United for a Better Tomorrow' in Kiruna in northern Sweden'.
>
> For two days, we discussed indigenous peoples' rights, oil drilling, and mining in the Arctic, as well as some of the challenges and threats that this amazing region is facing. Some twenty Arctic indigenous peoples organizations signed a joint statement calling for a ban on offshore oil drilling in the Arctic and a moratorium on onshore oil exploration in the region. The coalition included major Indigenous groups from all Arctic regions, two of which are permanent members of the Arctic Council.
>
> (Greenpeace Nordic 2013)

While Greenpeace lauded this meeting as a signal of Indigenous cooperation and Greenpeace's work in bringing about this work, the messaging was not without challenge. Greenpeace's anti-oil and mining messaging received immediate pushback from Inuit of Canada, with Terry Audla, president of Inuit Tapiriit Kanatami, reported to have said in the Nunatsiaq News (2013) that "I know that the legitimate Inuit claims organizations and leaders across Canada's Arctic regions do not share this view and we collectively reject Greenpeace's questionable use of the indigenous voice".

Part of the process of the Indigenous Peoples Policy is that Greenpeace offices are publicly committed to respect Indigenous sovereignty, self-determination, leadership, traditional activities, Indigenous rights and jurisdictions, efforts towards autonomy, international law, and, quite significantly for on-the-ground campaigning:

> Established processes within Indigenous Peoples' communities and commits to transparency regarding issues or activities that pertain to or would affect the local Indigenous communities. We seek input and guidance early in our planning process around events and/or actions that will take place on their traditional territories and spiritually significant sites.
>
> (Greenpeace USA n.d.)

While there may be principles in place, like the principle/philosophy of nonviolence, to help theoretically guide the organization and its accountability, there is an observable tension in Greenpeace's effort to address challenging elements of its past and current campaigning, and the attitudes and beliefs that underpin them. The reality is that much of Greenpeace's support base is unaffected by activism and policy interference and misrepresentation of seal hunting, as they have few, if any, personal connections to cultural and economic sealing practices or practitioners. It is not audiences in the Circumpolar North that are driving Greenpeace's donor base, but rather people predominantly in European and North American urban and suburban centres.

With its support base, Greenpeace must be care about treating it anti-sealing legacy as if it is problematic because to them the legacy is largely viewed as a period of triumph. As Outahsen reflected,

> one [Dutch] journalist that didn't really know of the whole history, which obviously a lot of people don't really know, right [came across Burgwald's media event with the seal skin vest]. Indigenous people know it and maybe some people that work there for the government, but a Dutch journalist there has no clue and thinks "What! They are in favour of fur![24]

Beyond the Anti-Sealing Campaign

The anti-sealing campaign demonstrated intolerance and colonialism, but it was not the only campaign by environmental and animal rights activists to target the Circumpolar North and negatively impact Indigenous peoples, just arguably the most impactful one. Other prominent campaigns that Greenpeace has been part of that have adversely impacted its relationships with various Indigenous peoples in the Circumpolar North include its anti-whaling work (Save the Whales) and its anti-oil and gas (Save the Arctic) campaigning.

This book will not go into Greenpeace's anti-whaling and anti-oil and gas work in great detail, but they are important to acknowledge as they link back to the discussion in Chapter 2 on Eurocentrism. These campaigns also point to issues repeated in this chapter by ISI, that issues of colonialism are present in the environmental and animal rights movement. As this chapter explored, Greenpeace has an uneasy relationship, trying to balance and address a legacy of colonialism in its activism while not wanting to completely disallow its activist history at the same time.

Greenpeace started its anti-whaling work on April 27, 1975 (Greenpeace Staff 2010), and its work centred on its "Save the Whales" slogan. Its worked to stop commercial whale hunting focused on campaigning against Japanese, Soviet, and Norwegian whalers (Zelko 2013). But its work to protect whales from commercial hunting caused issues for Indigenous and non-Indigenous cultures who practice traditional whale harvesting. Greenpeace adhered to similar communication techniques as its anti-sealing work.

Though Greenpeace's work against commercial whaling dates to the era of its anti-sealing work, it maintains public support for its prominent anti-whaling work and still communicates about it in a positive light

(Mackenzie 2018; Greenpeace USA 2025). Similar to its stance on sealing, Greenpeace has walked back its blanket opposition to the killing of whales to acknowledge Indigenous traditional hunting (e.g. Hopper 2014). In 2014, for example, Greenpeace showed support for the Inuit of Kangiqtugaapik/Clyde River (the relationship between Greenpeace and Kangiqtugaapik/Clyde River is discussed more in Chapter 5) and their first whale hunt in more than 100 years (Hopper 2014). But a consequence of Greenpeace's contemporary efforts to address its inadvertent impact on Indigenous hunting rights and cultural practices is that it now faces severe criticism from organizations like Sea Shepherd Conservation Society (Nunatsiaq News 1997); the organization formed by Paul Watson when he was expelled from Greenpeace in 1977 (Burke 2021, 2023). A side-effect of both the early effectiveness of Greenpeace's Save the Whales campaign and its more recent support for Inuit/Indigenous traditional whale hunting is that it faces headlines pointing to the inconsistence in its position over time, such as from the Canadian newspaper *National Post* in 2014 when it supported the Kangiqtugaapik/Clyde River hunt: "Save (most of) the whales: Greenpeace now supports Inuit hunting, but Native groups still wary" (Hopper 2014).

Unlike the seal issue, Greenpeace's anti-whaling work did not have the same level of blowback for the organization. This is likely because many whale species are endangered and were in need of regulatory intervention to protect them, such as rules around preventable instances of animal harm in shipping lanes with boat encounters with whales (Fisheries and Oceans Canada 2022). Furthermore, whale hunting and the use of commercial whale products have been in a natural decline and out of fashion in most parts of the world long before campaigning started because of the economic shift away from the use of things like blubber as fuel and bones in corsets. So, the degree of public resistance, even in areas where whales are hunted, to the need for some new levels of protection to ensure the preservation of whale populations has had, and continues to have, widespread support, even if there is a dislike towards Greenpeace and its communication style in its campaigning.

Even when the number of harp seals was low in the Arctic and sub-Arctic/North Atlantic and the grey seals, for example, were extremely low in the Baltic Sea region in the 1970s, the species situations were not comparable to those of some whale species. When the number of harp seals was presented by activists as endangered in the 1970s, for example, there were still 1.5–1.8 million animals off the coast of Canada alone (Fisheries and Oceans Canada 2012); a good number to ensure genetic diversity to aid with species recovery, which happened in the subsequent decades. In contrast, the fin whale, for example, which was on the brink of extinction in the 1970s, as of 2018 still has an estimated "less than 1,000 mature individuals" in Canadian waters (Fisheries and Oceans 2022).[25]

The point of this discussion is not to say that the work against whaling by Greenpeace and other animal rights groups has not been problematic. The across-the-board messaging to "Save the Whales" gives the impression of uniformity in the circumstances of all whale species, which is misleading. The Faroe Islands, for example, continues to be attacked by anti-whaling activists and their supporters, with Sea Shepherd being the most prominent organization (Burke 2022b, 40), despite the whales hunted there, the pilot whale, not being endangered. It is estimated, for example, that there are at least 380,000 long-finned pilot whales as of 2015 (NAMMCO 2021).

The 21st-century Save the Arctic campaign, however, lands a bit differently in terms of organizational legacy. Whereas the anti-sealing and anti-whaling campaigns may be viewed with some degree of generosity for being products of their time, in terms of what was considered acceptable in environmental and animal rights activism when the work started, the Save the Arctic campaign does not have the shield of history to hide behind. As such, the early years of the Save the Arctic campaign stood as a turning point for Greenpeace in the 21st century and appear to show the organization learning quickly that it needed to recalibrate its approach towards Indigenous peoples and its depiction of the Arctic.

Greenpeace states that its Save the Arctic campaign started in 2012 (Greenpeace Canada 2025a),[26] but there is ample evidence to indicate that it started – even if it was not formally called Save the Arctic yet – in 2010 in response to the Gulf of Mexico Deepwater Horizon drilling rig explosion. This connection is evident in Greenpeace's work in Kalaallit Nunaat/Greenland in 2010, shortly after the Deepwater Horizon incident, protesting Cairn Energy's oil exploration in Kalaallit Nunaat/Greenland "in the wake of BP's Gulf of Mexico oil spill disaster" (BBC News 2010; also see Carrell 2010).

The campaign is focused on banning oil exploration and development in the Arctic region; "the Save The Arctic project [aims] to protect the pristine Arctic from a possible oil spill" (Greenpeace Canada 2025a). Greenpeace introduces the campaign on its Greenpeace Canada website by stating,

> For millennia people lived in the Arctic alongside some of the most elusive and incredible animals, in some of the toughest conditions on Earth. But now all that is changing. The Arctic is heating up rapidly because of climate change.

While the wording of the Save the Arctic campaign's petition introduction acknowledges that people reside in the Arctic, the imagery often associated with that description by Greenpeace and other organizations in their Arctic work does not include people in its visual depictions or descriptions of the region. Instead, imagery of a lone polar bear on a floating piece of ice, for example, dominates (Greenpeace Canada 2025b).

A core issue with the Save the Arctic campaign is Greenpeace's playing upon and perpetuating the "pristine Arctic" trope to frame the region, incite a sense of urgency, and galvanize actors to support their agenda. Presenting and perceiving the Arctic as a pristine wilderness is well-established in the global imagination through film, song, literature, media, and national identities. Keskitalo (2004) notes that the idea of pristine-ness is tied to the concepts of wilderness and frontier, which are both rooted in Anglo-American cultural romantic conceptualization and imagining of history, time, and space.[27] The linking of the idea of pristine with wilderness is arguably "a twentieth-century phenomenon, born in North America following upon the growing conservation movement that established the world's first national parks" (Keskitalo 2004, 209).[28]

The romanticism of the Arctic, embodied by the marketing of the Arctic and the North as pristine by environmental activists, harkens to a colonial idealization of the region as otherworldly. As Keskitalo (2004, 36) points out about interest in Arctic cooperation started in the late 20th century:

> The foci on environmental protection and the concept of sustainable development characteristic of the 1980s came to combine with a romantic view of the Arctic derived from exploration. The image of the Arctic this also in recent descriptions largely remained that of a pristine wilderness. What was new was that statements often posed the romantic view of Arctic nature against the *threat* to it… "the Arctic" [is]…now re-interpreted as an area at risk.

A key problem with the pristine narrative is that it erases peoples and cultures in the Arctic, and the Circumpolar North more broadly. The Circumpolar North is a lived-in space with many different Indigenous peoples like Inuit, Sámi, Innu, Gwitch'in, and Aleut, as well as long-established non-Indigenous peoples, many of whom practice a hybrid Westernized lifestyle with traditional dimensions like many Icelanders, Faroese, Norwegians, Finnish, Newfoundlanders and Labradorians and the peoples of the Magdalen Islands (who can be both Indigenous and non-Indigenous), Ålanders, and the peoples of Svalbard.

Furthermore, when people are acknowledged by the same environmental and animal rights activists who present the Arctic and North as pristine and in need of protection, there is a natural conflict whereby people, through their very presence and desire to do things in and to their local area, become viewed as a problem. Svalbard, for example, has a history of being subject to equally misleading portrayals as "a virtually untouched environment" and as an area that has been subject to environmental "rape"

(Kruse 2016, 518). Kruse (2016, 518) argues that both sweeping narratives about Svalbard are "extremes [that] underrepresent and devalue centuries of human commitment and endeavour in the remote archipelago".

Indigenous peoples of the North run into similar issues as the people of Svalbard, wherein their existence and rights are approached by some activists as tolerable provided that they live "traditionally", though what exactly is meant by living traditionally is loosely conceived and seems to centre on the overarching impression that so long as they act as activists would like and do not ruin anything environmentally. This way of including Indigenous peoples is exactly how the sealing issue has been approached and framed by prominent anti-sealing organizations and the European Union; so long as Indigenous peoples only hunt seals in a very limited interpretation of subsistence traditional hunting, then they are seen as morally permissible to do so (Burke 2023).

The Save the Arctic campaign rubs against this tension between wanting to acknowledge Indigenous peoples and rights, with the parallel desire for an outright ban on all oil exploration in the region. While it may be true that some Indigenous peoples in the Arctic are opposed to some or all oil exploration in their homeland (e.g. discussed more in Chapters 5 and 6), the view that all are opposed or should be opposed is incorrect. In Kalaallit Nunaat/Greenland and Alaska, for example, there is public support for oil drilling and exploration, though the subject is divisive as people weigh the benefits of potential and established revenue streams from non-renewable resource development with the real and potential environmental and socio-economic impacts (Lydon 2024; Spitzer 2001).

In Kalaallit Nunaat/Greenland, for example, where Greenpeace launched its Save the Arctic protests in 2010, the idea of oil exploration and potential development is tied to domestic discussions about self-determination and independence from Denmark (Gad 2014; Burke 2020). As Ulrik Gad (2014, 105) argues,

> the substantial resources deemed necessary for Greenlandic sovereignty to be more than an empty shell appear to be materializing as the Arctic ice continues to melt – not just because the climate becomes less prohibitive for the extraction of Greenlandic minerals but also because the accompanying interest from new business partners and rising powers conveys the impression that Greenland has alternatives to Denmark.

In addition, Marc Jacobsen notes that "[t]he interest in oil and gas varies greatly from one Arctic country to another" (Jacobsen 2020, 186). The idea, therefore, that Kalaallit/Greenlandic Inuit should outright oppose all oil exploration and the potential economic and sovereignty benefits they might derive from it at the behest of Greenpeace activists or simply because they are Indigenous, is unrealistic and simplistic, especially given the organization's anti-sealing activism and how the success of the movement undercut the sustainable sealing industry which has hurt, and continues to hurt, many Kalaallit communities and families (Burke 2020).

Greenpeace's Save the Arctic work is ongoing. And as the subsequent chapters illustrate, there is not an outright rejection of the organization's work against oil exploration by Indigenous peoples in the Circumpolar North. The subject of non-renewable resource development is complicated, often contentious, and subject to change, but it is worth noting that its start in Kalaallit Nunaat/Greenland did not have the domestic resonance there that Greenpeace would have hoped for (Burke 2020; Carrell 2010). Offshore oil exploration by Cairn in the mid-2010s was not swayed by Greenpeace's work; rather a lack of evidence of sufficient viable oil deposits stopped further exploration at that time (Burke 2020).

Conclusion

Greenpeace has apologized, in part, for some of its anti-seal campaigning, and those efforts have caused some internal disquiet. Many members and supporters of the organization are proud and supportive of the organization's anti-sealing legacy and do not want backtracking on it. But the legacy of Greenpeace's anti-sealing campaigning is complicated.

As the subsequent chapters help illustrate, some activists within Greenpeace are looking to change how the organization involves itself in local issues in the Circumpolar North. This is not to stay that Greenpeace's anti-sealing legacy is forgotten, rather there appears to be some effort by northern-focused activists working on behalf of Greenpeace to try and not replicate past mistakes, and in at least one instance – outlined in Chapter 5 – address part of it head on as the inroads with Inuit in Canada are directly tied to the aforementioned 2014 apology.

Notes

1 There is an inconsistency about the exact formal start date of Greenpeace's entrance into anti-sealing campaigning. For example, Allen (1979) states that Greenpeace started its anti-sealing campaigning in 1975, whereas Joanna Kerr (2014), in the Greenpeace Canada apology to Inuit, Indigenous, and coastal peoples, states that Greenpeace started its anti-sealing campaigning in 1976.

2 Though 1969 is often attributed as the start of what became known as the anti-sealing movement (Patey 1990), Wenzel (1987, 199) reflects that in 1967, "a fresh round of protest over the Atlantic harp seal kill flared. It was at this time that a 'Save the Seals' campaign came into being, ultimately developing into the International Fund for Animal Welfare". Allen (1979, 425), too, notes that "[t]he modern anti sealing

industry dates from 1967, when IFAW was formed". IFAW formally started in 1969, developing from the 1967 Save the Seals campaign, so 1969 will continue to be used as the formal start of the anti-sealing movement.

3 Norwegian sealers were also targeted, though much of the on-the-ground campaigning against sealers in the 1960s–1970s focused on rural Canadian fishing communities in Newfoundland and Quebec.

4 Fishers/sealers who participate in the North Atlantic seal hunt also operate out of Nova Scotia and Prince Edward Island, though these areas have not been prominent in anti-sealing campaigning.

5 Europe was the primary market for seal product exports from Canada in the 1980s, and the implementation of the seal product import ban in 1983, followed by a complete ban on seal product imports with some minor exceptions for Indigenous products in 2009, effectively crippled the traditional sealing industry (European Commission 2023).

6 Burke writes of this time period that

> Rural-based sealers were disadvantaged in the fight, caught unaware, and unprotected while trying to practice traditional lifestyles. Inuit and other Indigenous hunters (such as the Innu and Mi'kmaq of Newfoundland and Labrador) got caught in the attack by actors that did not bother to learn the nuances of the societies they were targeting or how, and to what extent, they participated in traditional sealing practices and the sealing economy.
>
> (Burke 2020, 3)

7 Please note that Wilson's affiliation with Greenpeace in the text denotes her affiliation at the time of our interview.

8 Interview by Danita Catherine Burke with Jessica Wilson of Greenpeace Canada, October 30, 2018.

9 Monkeywrenching is ecological sabotage, also referred to as "ecotage" that entails "stopping environmentally destructive activities by damaging the equipment involved in the activity…to the point where the companies involved would no longer find those activities profitable enough to pursue them" (Vail and Goode 2008, 192). Monkeywrenching is presented as non-violent resistance because the actions or activities are "not directed toward harming human beings or other forms of life" (Doyle 2000, 46) but results in the use of tactics like spiking roads and trees and tampering with crucial bolts in machinery that can cause physical and psychological harm or death, for example, by leading to life threatening and life altering injuries to workers when they use tampered with equipment or machinery that is no longer safe to use (Goodin 1992, 134).

10 In a 1977 report by Greenpeace, the organization acknowledged that many of the people participating in seal harvesting had "3.5 dependents, [and] an average education of grade 9, living in isolated communities with limited occupational mobility" (Greenpeace Foundation 1977, 2). The organization was aware that the people they were targeting were isolated, had few alternatives, and, while experts in their field

(fishing, sealing), they lacked formalized education credentials that would open economic alternatives for them in the open labour market.

11 Permission was given by the individual from Greenpeace I interviewed to use their name in reference to quotes from the interview. However, given the nature of the quote and out of caution for the individual given the focus of the quote, I have decided to not give direct attribution. The quote is part of a reply to the question: "To what extent do you think that Greenpeace is having to live down the legacy of former members like Paul Watson who arguably went against core organisational values but in the organisation's name while a member?" and a broader discussion that was had on the topic of navigating Greenpeace's legacy and reputation in the northern spaces this individual operated.

12 The European Economic Community's voluntary ban on the import of seal products was first proposed in 1982 and came into effect in 1983 (Wenzel 1987, 199).

13 Indigenous Survival International "was established 1984 in Yellowknife, NWT, as an Canadian-Alaskan-Greenlandic partnership to protect Indigenous subsistence rights and protest the anti-harvesting movement. ISI was very involved in discussions of the World Conservation Strategy. The organization was dissolved in 2015" (Inuit Circumpolar Council Canada Archives 2025).

14 At the time of negotiations, what we now know as the World Wide Fund for Nature was then called the World Wildlife Fund.

15 The work by Indigenous Survival International references "Greenpeace (Denmark)" with the reference implying that Greenpeace (Denmark) is an independent national office. While this may have been the case in the 1980s, today Greenpeace's Danish office is part of its overarching Greenpeace (Nordic) organization.

16 Greenpeace's Danish office, as with all of Greenpeace, however, remains opposed to commercial-scale hunting of seals regardless of the ethnicity of the hunter (Burke 2020, 2024).

17 Interview by Danita Catherine Burke with Jessica Wilson of Greenpeace International, October 30, 2018.

18 Interview by Danita Catherine Burke with Mads Flarup Christensen of Greenpeace Nordic, February 7, 2019.

19 Interview by Danita Catherine Burke with Jessica Wilson of Greenpeace International, October 30, 2018.

20 Interview by Danita Catherine Burke with Mads Flarup Christensen of Greenpeace Nordic, February 7, 2019.

21 Interview by Danita Catherine Burke with Faiza Outahsen of Greenpeace Netherlands, September 11, 2018.

22 Interview by Danita Catherine Burke with Faiza Outahsen of Greenpeace Netherlands, September 11, 2018.

23 Interview by Danita Catherine Burke with Mads Flarup Christensen of Greenpeace Nordic, February 7, 2019.

24 Interview by Danita Catherine Burke with Faiza Outahsen of Greenpeace Netherlands, September 11, 2018.

25 Fisheries and Oceans Canada (2022) estimates that the number of mature fin whales: "do not include Canadian waters beyond the continental shelf where substantial numbers were sighted in a 2018 survey".

26 Greenpeace Canada states that the Save the Arctic campaign was the starting point of a broader campaign called "The People vs. Oil", started in 2018 (Greenpeace Canada 2025a).

27 In my previous research on the evolution of Canada's relationship with the Arctic and North, for example, I argue that:

> The narratives that encourage the view of the Arctic as a pristine area with a glorious and rich British naval history characterize the predominant romantic ideas. The belief in Canadian society that the Arctic belongs to Canada is driven by emotion and an emotional attachment to the myth of the pristine wilderness which the Arctic region embodies.
>
> (Burke 2018, 36)

28 Keskitalo (2004, 209–210) also observes that:

> [H]istorically and with effects into the present, the frontier myth this resulted in a separation of "wilderness" and "city," and into specific roles for the populations of such "wildernesses": as indigenous, they became conceived of as different and through their relation to the environment. As the wilderness was not a place for non-indigenous everyday life until it had lost its frontier role.

References

Allen, J. 1979. "Anti-Sealing as an Industry." *Journal of Political Economy* 87(2): 423–428.

BBC News. 2010. "Cairn Energy Strikes Oil off Greenland." September 21. https://www.bbc.com/news/uk-scotland-11381281

Bernauer, W. 2020. "Producing Consent: How Environmental Assessment Enabled Oil and Gas Extraction in the Qikiqtani Region of Nunavut." *The Canadian Geographer* 64(3): 489–501.

Bharadwaj, L.K. 1998. "Principled versus Pragmatic Nonviolence." *Peace Review* 10(1): 79–81.

Burke, Danita Catherine. 2018. *International Disputes and Cultural Ideas in the Canadian Arctic: Arctic Sovereignty in the National Consciousness*. Palgrave Macmillan.

Burke, Danita Catherine. 2020. "Re-Establishing Legitimacy after Stigmatization: Greenpeace in the North American North." *Polar Record* 56(e26): 1–12.

Burke, Danita Catherine. 2021. "The Case for a Greenpeace Apology to Newfoundland and Labrador." *The Northern Review* 51(1): 173–187.

Burke, Danita Catherine. 2022a. "Interview: Reflecting on the Seal Hunt and the Anti-Sealing Movement–A Conversation with former President of the Canadian Sealers Association James Winter". *Women in the Arctic and*

Antarctic, May 31. https://womeninthearcticandantarctic.ca/2022/05/31/interview-reflecting-on-the-seal-hunt-and-the-anti-sealing-movement-a-conversation-with-former-president-of-the-canadian-sealers-association-james-winter/

Burke, Danita Catherine. 2022b. *WWF and Arctic environmentalism: Conservation and the ENGO in the Circumpolar North*. Manchester University Press.

Burke, Danita Catherine. 2023. *Cultural Violence, Stigma and the Legacy of the Anti-Sealing Movement*. Routledge.

Burke, Danita Catherine. 2024. "Nonviolence and Greenpeace: Apologies and Anti-Sealing Campaigning." *Arctic* 77(3): 246–260.

Carrell, Severin. 2010. "Greenland's Prime Minister Lambasts Greenpeace for Raiding Arctic Oil Rig." *The Guardian*, August 31. https://www.theguardian.com/environment/2010/aug/31/greenland-greenpeace-arctic-oil-rig

Carter, Neil. 2007. *The Politics of the Environment: Ideas, Activism, Policy*. 2nd Edition. Cambridge University Press.

CBC. 1978. "Barbara Frum Interview with Paul Watson." https://www.youtube.com/watch?v=H_gTBDFTXE0

CBC Radio. 1965. "New Rules to Protect Seals." April 2. https://www.cbc.ca/archives/entry/new-rules-to-protect-seals

CBC The Broadcast. 2021. "Still Waiting for an Apology from Greenpeace." November 11. https://www.cbc.ca/listen/live-radio/1-122-the-broadcast/clip/15878064-still-waiting-apology-greenpeace

Crown-Indigenous Relations and Northern Affairs Canada. 2019. "Nunavut Lands and Resource Devolution Agreement in Principle." August 15. https://www.rcaanc-cirnac.gc.ca/eng/1565186303207/1565186324673

Crown-Indigenous Relations and Northern Affairs Canada. 2022. "Truth and Reconciliation Commission of Canada." https://www.rcaanc-cirnac.gc.ca/eng/1450124405592/1529106060525

Dale, S. 1996. *McLuhan's Children: The Greenpeace Message and the Media*. Between the Lines.

Davies, Brian. 1970. *Savage Luxury: The Slaughter of the Baby Seals*. Souvenir Press.

Doyle, T. 2000. *Green Power: The environment movement in Australia*. University of New South Wales Press Ltd.

Erickson, B. 2020. "Anthropocene Futures: Linking Colonialism and Environmentalism in an Age of Crisis." *EPD: Society and Space* 38(1): 111–128.

Essemlali, Lamya and Paul Watson. 2013. *Captain Paul Watson: Interview with a Pirate*. Firefly Books Ltd.

European Commission. 2023. "Report from the Commission to the European Parliament and the Council on the implementation of the Regulation (EC) No 1007/2009, as amended by Regulation (EU) 2015/1775, on Trade in Seal Products. Brussels, 19.10.2023 COM(2023) 633 final." https://eur-lex.europa.eu/legal-content/EN/TXT/?uri=CELEX%3A52023DC0633&qid=1698048882484

European Commission. n.d.. "Trade in Seal Products. Energy, Climate Change, Environment." https://environment.ec.europa.eu/topics/nature-and-biodiversity/trade-seal-products_en#:~:text=In%201983%2C%20the%20EU%20adopted,commercially%20imported%20into%20their%20territory

Fenge, Terry. 1985. Letter to the Editor of the Guardian Newspapers Ltd from Terry Fenge, Director, Policy Studies. Laurier Archives, Canadian Arctic Resources Committee fonds S469, Box 270, File 13.1.21(1).

Fisheries and Oceans Canada. 2012. "Limit Reference Points and Minimum Populations of Harp Seals (Pagophilus groenlandicus)." Canadian Science Advisory Secretariat Science Advisory Report 2012/067. https://publications.gc.ca/collections/collection_2013/mpo-dfo/Fs70-6-2012-067-eng.pdf

Fisheries and Oceans Canada. 2022. "Fin Whale, Pacific Population." *Aquatic Species at Risk Program, Pacific Region*. https://wildlife-species.canada.ca/species-risk-registry/virtual_sara/files/public/Cd-FinWhaleRorqualCommun-v00-2022Oct-Eng.pdf

Fur Institute of Canada. 1988. "A Strategy to Support Canada's Fur Trade: A Report to Federal/Provincial/Territorial Wildlife Ministers." Laurier Archives, Canadian Arctic Resources Committee fonds S469, Box 292, File 13.1.24.

Gad, Ulrik P. 2014. "Greenland: A Post-Danish Sovereign Nation State in the Making." *Cooperation and Conflict* 49(1): 98–118.

Gilday, Cindy. 1986. Letter to Benjamin P. Nageak, Department of Wildlife Management, North Slope Borough, Alaska from Cindy Gilday for ISI International. September 10, 1986. Laurier Archives, Canadian Arctic Resources Committee fonds S469, Box 48(3), File 2.9.3.

Goodin, R. 1992. *Green Political Theory*. Polity Press.

Greenpeace Canada. 1976. "Shepard's of the Labrador Front." *Greenpeace Chronicles* 2: 6.

Greenpeace Canada. 2017. "Greenpeace Canada Policy on Indigenous Rights." https://www.greenpeace.org/canada/en/policy-on-indigenous-rights/

Greenpeace Canada. 2025a. "People vs Oil." https://www.greenpeace.org/international/campaigns/people-vs-oil/

Greenpeace Canada. 2025b. "Save the Arctic." https://www.greenpeace.org/canada/en/save-the-arctic/

Greenpeace Foundation. 1977. "The Canadian Seal Hunt." Memorial University of Newfoundland, Queen Elizabeth 2nd Library, Centre for Newfoundland Studies File SH363.G711977.

Greenpeace Nordic. 2013. "Greenpeace Annual Report 2013." https://www.greenpeace.org/static/planet4-finland-stateless/2023/01/42366f36-greenpeace_annual_report_2013.pdf

Greenpeace Staff. 2010. "'Save the Whales' 35th Anniversary." *Greenpeace USA*, April 27. https://www.greenpeace.org/usa/save-the-whales-35th-anniversary/

Greenpeace USA. 2025. "Save the Whales." https://www.greenpeace.org/usa/oceans/save-the-whales/

Greenpeace USA. n.d. "Greenpeace USA Indigenous Peoples Policy." https://www.greenpeace.org/usa/about/greenpeace-usa-indigenous-peoples-policy/

Harter, J. 2004. "Environmental Justice for Whom? Class, New Social Movements, and the Environment: A Case Study of Greenpeace Canada, 1971–2000." *Labour/Le Travail* 54(1): 83–119.

Hawkins, R. and J.J. Silver. 2017. "From Selfie to #sealfie: Nature 2.0 and the Digital Cultural Politics of an Internationally Contested Resource." *Geoforum* 79(2017): 114–123.

Hofverberg, E. 2019. "Greenland's National Day, the Home Rule Act (1979), and the Act on Self-Government (2009)." *The Library of Congress*. https://blogs.loc.gov/law/2019/06/greenlands-national-day-the-home-rule-act-1979-and-the-act-on-self-government-2009/

Hopper, Tristin. 2014. "Save (Most of) the Whales: Greenpeace Now Supports Inuit Hunting, but Native Groups Still Wary." *National Post*, August 15. https://nationalpost.com/news/canada/save-most-of-the-whales-greenpeace-now-supports-inuit-hunting-but-native-groups-still-wary

Hummel, Monte. 1986. "Statement by Monte Hummel, President World Wildlife Fund Canada for the Aboriginal Peoples Conference on Harvesting and Renewable Resources." World Wildlife Fund Canada, Chisasibi, Quebec, January 30–February 1, 1986. Laurier Archives, Canadian Arctic Resources Committee fonds S469, Box 48(3), File 2.9.3.

Îles de la Madeleine. 2021. *Éphémérides: Un film choc tourné en 1964, Les grands phoques de la banquise*. https://www.ilesdelamadeleine.com/2021/05/ephemerides-un-film-choc-tourne-en-1964-les-grands-phoques-de-la-banquise-5/

Indigenous Survival International. 1985. "Indigenous Survival International (ISI): Submission to the Canadian Royal Commission on Seals and the Sealing Industry". London, England. April 9, 1985. Laurier Archives, Canadian Arctic Resources Committee Fonds S469, Box 48(1), File 2.9.3.

Inuit Circumpolar Council Canada Archives. 2025. "Indigenous Survival International." https://iccc-archives.andornot.net/link/authority12886

ISI Canadian Sub-Committee Meeting. 1985. Assembly of First Nations Offices, Ottawa, Ontario. April 4, 1985. Laurier Archives, Canadian Arctic Resources Committee fonds S469, Box 48(2), File 2.9.3.

Jacobsen, Marc. 2020. "Greenland's arctic advantage: Articulations, Acts and Appearances of Sovereignty Games." *Cooperation and Conflict* 55(2): 170–192.

Kalland, A. 2009. *Unveiling the Whale: Discourses on Whales and Whaling*. Berghan Books.

Kerr, Johanna. 2014. "Greenpeace Apology to Inuit for Impacts of Seal Campaign." *Greenpeace Canada*, June 24. https://www.greenpeace.org/canada/en/story/5473/greenpeace-apology-to-inuit-for-impacts-of-seal-campaign/

Keskitalo, E. C. H. 2004. *Negotiating the Arctic. The Construction of an International Region*. New York and London: Routledge.

Kruse, Frigga. 2016. "Is Svalbard a Pristine Ecosystem? Reconstructing 420 Years of Human Presence in an Arctic Archipelago." *Polar Record* 52(2006): 518–534.

Kuokkanen, R. 2017. "The Pursuit of Inuit Sovereignty in Greenland." *Northern Public Affairs* 5(2): 45–48.

Laurier Archives CARC holdings. 1985. "Brown Peace Browned Off at Greenpeace Peace". Press Release. January 15 and 16, 1985. Laurier Archives, Canadian Arctic Resources Committee fonds S469, Box 48(1), File 2.9.3.

Loukacheva, N. 2012. "Indigenous Inuit Law, 'Western Law' and Northern Issues." *Arctic Review on Law and Politics* 3(2): 200–217.

Lydon, Tim. 2024. "Alaska Governor Awards $1m in State Funds to Indigenous Group Backing Oil Drilling." *The Guardian*, October 31. https://www.theguardian.com/us-news/2024/oct/31/alaska-indigenous-oil-drilling

Mackenzie, Willie. 2018. "A Brief History of Commercial Whaling and Greenpeace." *Greenpeace International*, September 10. https://www.greenpeace.org/international/story/18307/history-commercial-whaling-greenpeace/

NAMMCO. 2021. "Long-Finned Pilot Whale." https://nammco.no/long-finned-pilot-whale/

Nunatsiaq News. 1997. "Greenpeace under Attack for Helping Inuit Whalers." October 10, 1997. https://nunatsiaq.com/stories/article/greenpeace_under_attack_for_helping_inuit_whalers/

Nunatsiaq News. 2013. "Indigenous Statement Calls for Arctic oil Development Moratorium." https://nunatsiaq.com/stories/article/65674 indigenous_statement_calls_for_ban_on_all_arctic_oil_development/

Partelow, S., K.J. Winkler and G.M. Thaler 2020. "Environmental Non-Governmental Organizations and Global Environmental Discourse." *PLoS ONE* 15(5): e0232945.

Patey, Francis. 1990. *A Battle Lost: An Unsuccessful Attempt to Save the Seal Hunt*. Robinson-Blackmore Printing and Publishing Ltd.

Phelps Bondaroff, Teale N. and Danita C. Burke. 2014. "Bridging Troubled Waters: History as Political Opportunity Structure." *Journal of Civil Society* 10(2): 165–183.

Phillips, L. 2010. "Inuit sue EU over Seal Ban." *EU Observer*, January 15. https://euobserver.com/green-economy/29273

Qikiqtani Inuit Association. 2013. "Qikiqtani Truth Commission: The Official Mind of Canadian Colonialism." https://www.qtcommission.ca/sites/default/files/public/thematic_reports/thematic_reports_english_official_mind.pdf

Rodon, T. 2017. "Offshore Development and Inuit Rights in Inuit Nunangat." In *Governance of Arctic Offshore Oil and Gas*, edited by Cécile Pelaudeix and Ellen Margrethe Basse. Routledge.

Roswell, Harry. 1978. "1978 Sealing Activities with Harp and Hooded Seals on the Front: A Report to the Committee on Seals and Sealing." Centre for Newfoundland Studies, Queen Elizabeth 2nd Library, Memorial University of Newfoundland and Labrador, St. John's, Canada. Stacks SH363 R691978.

Roswell, Harry C. 1977. "1977 Sealing Activities by Newfoundland Landsmen and Ships on the Front: A Report to the Committee on Seals and Sealing and the Canadian Federation of Humane Societies." Centre for Newfoundland Studies, Queen Elizabeth 2nd Library, Memorial University of Newfoundland and Labrador, St. John's, Canada. Stacks SH363.R69.

Sharp, Gene. 2005. *Waging Nonviolent Struggle*. Porter Sargent.

Sinclair, Peter, Robert Hill, Cynthia Lamson and H.A. Williamson. 1989. "Social and Cultural Aspects of Sealing in Atlantic Canada." Report for the Royal Commission on Seals and the Sealing Industry in Canada. Institute of Social and Economic Research, Memorial University of Newfoundland.

Spitzer, Aaron. 2001. "Alaskan Inuit Support Oil Drilling in the ANWR." *Nunatsiaq News*, January 19. https://nunatsiaq.com/stories/article/alaskan_inuit_support_oil_drilling_in_the_anwr/.

Svels, Kristina, Pekka Salmi, Nelson F. Coelho, Viktor Eriksson, Sara Königson, Esa Lehtonen, Sven-Gunnar Lunneryd, Petri Suuronen, Markus Vetemaa and Åsa Waldo. 2025. "Struggling towards Co-Existence of the Baltic Sea Coastal Fisheries and the Grey Seal." *Maritime Studies* 24(1): 1–15.

Thoener, Victorine Che. 2024. "How Indigenous Peoples and Local Communities Can Protect and Restore Biodiversity." https://www.greenpeace.org/international/story/70942/indigenous-people-local-communities-protection-biodiversity-climate/

Vail, D.A. and E. Goode. 2008. *Extreme Deviance*. SAGE Publications Inc.

Van Holstein, E. and L. Head. 2018. "Shifting Settler-Colonial Discourses of Environmentalism: Representations of Indigeneity and Migration in Australian Conservation." *Geoforum* 94(1): 41–52.

Wenzel, George. 1987. ""I Was Once Independent": The Southern Seal Protest and Inuit." *Anthropologica* XXIX(1987): 195–210.

Weyler, Rex. 2004. *Greenpeace: How a Group of Ecologists, Journalists, and Visionaries Changed the World*. Raincoast Books.

Woods, S.J. 1986. "The Wolf at the Door: The Anti-Harvest Campaign Strikes at the Heart of North Aboriginal Economies." *Northern Perspectives* 14(1): 1–8.

Zelko, Frank. 2013. *Make It a Green Peace! The Rise of Countercultural Environmentalism*. Oxford University Press.

4 Traditional Lands and Reindeer Herders

Sámi, Finland

The Sámi[1] peoples of Sápmi[2] are the Indigenous people in Northern Europe with a deep-rooted cultural and economic connection to the practice of reindeer husbandry (Borchert 2001, 1; Nyhlén et al. 2018, 81). That said, not all Sámi are reindeer herders, and there are internal complexities with Sámi and non-Sámi societies across the Nordic region regarding perspectives on the use of territory for reindeer herding and other interests (Cederberg 2010).[3] The issue of land access and rights of Sámi, and their reindeer herders especially, is a major point of contention in the Nordic[4] countries, particularly as the issue of Indigenous land access and rights for traditional practices can conflict with resource development and infrastructure proposals and ambitions. This issue is exasperated by the reality that "[d]efining Saami traditional lands and waters on a local level is not an easy task" (Wheelersburg 2008, 163).

One of Greenpeace's earliest, high-profile, and arguably successful engagements in the Circumpolar North in the 21st century was in Finland, working with Sámi reindeer herders. Chapter 4 introduces some of the elements of this campaigning in Finland, specifically focusing on how Greenpeace approached the issue and local peoples, and why this campaign stands out regionally for being innovative with a locally driven approach in contrast to the past anti-sealing campaign. This chapter illustrates that Greenpeace's work in Finland signified the benefits of a bottom-up approach when entering into a new area – both geographically and topically. The chapter also highlights the power of developing positive relationships with local stakeholders and rightsholders when work is locally driven and instigated with attention given to the cultural and economic significance of the issue, even when it may have some incongruence with organizational and membership views on wider issues such as human use and killing of animals.

Greenpeace, Sámi of Finland, and the Forestry Industry

The issue of land rights and access was particularly pronounced in the 1990s and early 2000s in Finland. A recurring problem in the Nordic

DOI: 10.4324/9781003674023-4

countries is the failure to acknowledge and respect Sámi land rights and access. In 2023, for example, the Norwegian government apologized to Sámi reindeer herders on the Fosen peninsula for issuing licences to build and operate wind farms in 2013 in an area known to be of significant cultural and economic importance to the Sámi. Specifically, the apology acknowledges that decisions by the Norwegian government "are a violation of human rights". The apology for the approach to the wind farms licensing on the Fosen peninsula "is a response to the Supreme Court's finding that the licensing decisions made in 2013 did not include satisfactory mitigating measures to prevent a violation of human rights" (Ministry of Petroleum and Energy 2023).

In Finland, for example, Sámi have faced considerable pressure over proposed plans – currently dormant – for the Arctic Railway Initiative, which was proposed in 2017 (Quinn 2019; Kuokkanen 2024); the Arctic Railway Initiative was a joint Finnish–Norwegian idea. The plan was for a railway that "would stretch about 500 kilometers from Finland's existing rail network in Rovaniemi to the Arctic Ocean at Kirkenes, a Norwegian port on the Barents Sea" (Nilsen 2021). This plan would have put the railway through the Sámi homeland, but consultations between state governments and Sámi were dismissed by Sámi as inadequate (Cepinskyte 2018). Kuokkanen reported that "[t]he Sámi Parliament was not included in the initial consultations despite its multiple requests for consultations with the Minister of Transport and Communications" (Kuokkanen 2024, 159).

As for reindeer herders, Else Grete Broderstad (2022, 16), in their work on Indigenous rights, international law and wind power development, states: "Those living off the land, like the reindeer herders, face major challenges in adapting and responding to different exploitation and development projects such as mining, power lines and hydropower plants, cabin sites, roads, railroads, and … wind power". Reindeer herders in Finland faced significant pressure from the logging industry, which sought to harvest forests that had long been used by herders for the feeding and movement of their reindeer.[5] A chief concern of reindeer herders with the forestry industry and its tree harvesting is that the "harvesting, soil scarification, road construction and other forestry-related activities diminish, deteriorate and fragment lichen grounds, cause additional work for herders and further decrease the possibility for reindeer to graze freely" (Lawrence and Raitio 2006, 2).

Finnish law recognizes the rights of Sámi to practice their culture, and the "Finnish government has been criticised…for not ensuring adequate practical protection of reindeer herding" (Lawrence and Raitio 2006, 1; also see Library of Congress 2021).[6] This perceived disconnect between the law and practice in Sámi-Finnish relations ultimately centres on a debate about Sámi inclusion and consent as it pertains to issues that impact their traditional lands, cultures, practices, and peoples. Catherine

Howlett and Rebecca Lawrence point out, in the context of reindeer herding, "for many Sámi communities, the ultimate goal is not dialogue, but to protect remaining reindeer grazing lands from further industrial encroachments" (Howlett and Lawrence 2022, 153).

According to Dubash and Oppenheimer (1993, 267), in their work on modifying the mandate of institutions, "NGO activities are typically stimulated by an innovative idea, a specific policy challenge to be met, a particular opportunity to be exploited or some combination of these". Greenpeace entered the picture of Nordic state relations with Sámi reindeer herders in Finland in 2002.[7] That year, Greenpeace Nordic's Finnish office started what would become an eight-year alliance with Sámi reindeer herders in a joint effort to preserve traditional grazing areas in the Sámi traditional homeland, Sapmi, in Northern Finland (Nyyssönen 2022).

In the municipality of Inari, a focal area for the dispute, Jukka Nyyssönen (2022, 2) argues that the land is 90 per cent owned by the state, which is a point of debate on its own between the Finnish government and the Sámi. The use of the government-owned land is

> divided as follows: (1) land area in economic usage; (2) wilderness areas, with limited zones for forestry usage; and (3) conserved areas, in which forestry is totally forbidden. The disputed areas belong mostly to category 1, but there were also plans to log within the wilderness areas.

Reindeer herding is permitted in the whole region, "but the pasture area for each reindeer-herding cooperative is delimited and the stress from forestry is unevenly shared between the cooperatives" (Nyyssönen 2022, 2).[8]

When Greenpeace entered the picture in 2002 at the invitation of some concerned Sámi, they worked to co-develop a strategy to target the logging industry to protect old-growth forests, focusing on the Finnish government forest enterprise Metsähallitus.[9] The reason why the issue of deforestation is so crucial to Sámi reindeer herders in Finland is that "[i]n Finland and in some parts of Sweden, reindeer herds remain primarily within the same forest range throughout the year, but summer and winter grazing usually take place in different areas within the range" (Vuojala-Magga et al. 2011, 228).[10] At this time Greenpeace's desire for the protection of old-growth forests aligned in many respects with Sámi environmental concerns: "The Sámi gaze upon nature conservation was a gaze to a preserved and saved resource zone; the questions of nature conservation were issues of compensations and securing usage rights" but notably "did not mean abandoning the cultural or economic side of the issue" (Nyyssönen 2022, 5).

In the end, the campaign ran by the ally-ship between Greenpeace and Sámi organizations appears to have been received as generally successful for both Greenpeace and Sámi[11]; "Almost 80% of the forests defined as important by reindeer herders and Greenpeace in 2002 have now been set outside of forestry either permanently or for the next 20 years" (News. Cision 2010; also see Cultural Survival 2015). Reportedly, when the campaign started, "[f]or several years, the Finnish government was unwilling to seek resolution to the conflict" but the turning point in the protests occurred when "in 2006 … [the] Finnish paper company StoraEnso responded to the campaign and decided to stop buying wood from the disputed areas" (News.Cision 2010).

From the vantage point of examining Greenpeace's evolution and how it approached this campaign, there are two big takeaway points. First, Greenpeace does not appear to have involved itself in local issues until it was approached by concerned Sámi. This approach then led to Greenpeace reaching out and forming working relationships with organizations that were comprised of local rightsholders and stakeholders, peoples most directly affected by the environmental problem under dispute. Second, Greenpeace's role in the campaign was primarily to help elevate and foster awareness of local concerns: loss of traditional forested grazing areas, including old-growth trees.

Sini Harkki is a member of Greenpeace Nordic[12] who has worked extensively on the Finnish forest campaigns. According to Harkki, their recollection, Greenpeace and Sámi decided to focus on old-growth forests in their work.

> I started with the forest campaign, assisting with the writing of some reports for Greenpeace International that had to do with Finland basically. And then I worked with the forest campaign mainly protecting the old growth forest of North Finland, also we did a lot of cooperation with the Indigenous people of Lapland – the Sámi – especially reindeer herders. That was a big part of my first years. It was devoted to that.[13]

Harkki provides some interesting insight in their interview on their views on how the alliance between Greenpeace Nordic and the Sámi worked in the context of the forest campaign.

For starters, in Harkki's opinion, there were two key things that, from Greenpeace's perspective that effectively kick-started the cooperation with Sámi in 2002.

> It started with two things happening at the same time. Two young Sámi activists basically walked into our office and said, "can you do something to help? We have a problem with industrial logging and they are not listening to the reindeer herders." And then at the same time one of the long term Finnish forest activists and campaigners who had moved

> to the northernmost part of Lapland to study to be a carpenter started talking with Sámi reindeer herders and they actually said "we need help and do you think you could help?" And then he got in contact with us. We got it [message to assist] from two directions from the Sámi themselves that they need some help campaigning for the forest and then that's how we started building it [the campaign to protect the forest] up. I think it took like two years of preparation that we decided that we need to do. It was maybe a different time, it was like 2003–2004. Nobody was really talking about principles like free and prior informed consent. It was quite primitive in the sense of understanding Indigenous rights. We, of course, knew about them, but it was not like Finland was ready to campaign and win campaigns with that theme. So the way we started working about it was that we went around all the Sámi reindeer cooperatives and asked them to basically draw on a map where their most important reindeer herding areas are and where they think forests should be preserved.[14]

It appears, therefore, that the instigation of Greenpeace's involvement in forest conservation and Indigenous rights issues in Northern Finland was prompted by members of the Finnish Sámi community. Only after local instigation did Greenpeace plan to get involved in helping to address forestry concerns, and the plan was built from the ground up with locals; at least this is Greenpeace Nordic's perception in Finland of how they became involved and how their campaign was created.

According to Mads Flarup Christensen, Executive Director of Greenpeace Nordic,[15] Greenpeace understood that for any campaign it became involved in on northern Finnish forestry issues in Sapmi to be viewed as legitimate, it was vital that it had a well-defined and publicly acknowledged working relationship with local Indigenous peoples. Christensen stated,

> Our approach there … from the beginning was to partner up and do what we do in alliance with Indigenous organisations. I think we [Greenpeace] Nordic have done that, since approximately maybe the late 1990s, the beginning of the 2000s when we worked with the Sámi people in the Northern forest issues [in Finland] and we did that in Sweden as well.[16]

The approach presented by Christensen, if it transpired as the reflections outlined suggest, illustrated a strategic level of self-awareness from members of Greenpeace leadership in the Nordic offices about who they needed as an ally, broadly speaking. Ally-ship was vital to legitimize Greenpeace's involvement with local audiences who would be responsible for implementing, complying, and living with any outcome that

Greenpeace advocacy might help achieve or cause. The move demonstrated that at least someone in a leadership position with internal sway at Greenpeace Nordic or Greenpeace International knew that the organization needed to situate itself within the then-growing trend of close cooperation with Indigenous peoples happening in the global environmentalist movement and push successfully for the organization to move in this direction.

Alternatively, Greenpeace Nordic may have stumbled upon making the right decision at the right time, with positive results, and some members of the organization realized that this approach to northern/Arctic engagement was a potential winning formula. Either way, the more effective long-term decision for Greenpeace was determined to be distancing its campaigning style and agenda from the "colonial residues" in the environmental movement, as Jukka Nyyssönen phrased it (Nyyssönen 2022, 5). This change meant that Greenpeace needed to ensure that its work on the issue of old-growth forest conservation in Northern Finland not only included the Sámi but also actively included them in a leadership role for the formulation, implementation, and negotiation of the campaign and any proposed outcomes.

We can see, therefore, based on Harkki's reflections on the campaign, that from the beginning, Greenpeace appears to have framed its involvement in the debate about forest conservation and reindeer herder rights as being driven by its alliances with local rightsholders and stakeholders. As Harkki remembered,

> Our main allies [during the forest protection campaign against the forestry industry] were the Sámi cooperatives, but later we also worked with the Sami Council and we coordinate a lot with the Sámi Parliament, but that campaigning was mainly with the reindeering cooperatives.[17] That's how it was organised.[18]

Harkki elaborated that

> Reindeer herders are organised in Finland which is actually something that has been imposed by the state of Finland that they need to be organised into these cooperatives that are these business units and geographical units as well. [*Interview: So you went to the cooperatives?*] Yeah, exactly and they all have a chair person and they all have board meetings where we went to establish these set of maps which was our common goal to work toward that these forests need to be protected. Our argumentation was like those [are] very important for nature though they were not all open forests and then on the rights of traditional Sámi reindeer herding. Then we started first working nationally. At the time the reindeer herders have very few people that were familiar

with any kind of advocacy work or had any time for that so we organised trips to meet civil servants and politicians and meet media and introducing this issue more nationally. We spent a couple of years doing that, so adding some NGO pressure on the state forest companies who were doing the logging. When that didn't really work, in 2005 we decided that we needed to do a more international campaign. At the time I was a freelancer for Greenpeace, but I was involved in all of these stages and then we started to develop the campaign together with the Sámi reindeer herders, thinking what it could be and it ended up being kind of a forest rescue station camp for a couple of months in the Sámi area … we brought media in and did activities with the Sami and of course campaigned in the markets for the forest Finnish industry.

Greenpeace's Northern Finland work was situated as part of its broader global advocacy on the issue of old-growth forest protection, and this framing of its advocacy gained traction.

The organization was successful, for example, in getting major support for companies like Ikea to "end all purchases of furniture made using wood from ancient forests" in 2010 (Greenpeace USA 2010). Boycotts like Ikea's aided the shared Greenpeace– Sámi in Finland agenda of forest conservation. However, before such an outcome was possible, the Greenpeace and Sámi relationship had to develop, and this had to happen parallel to the advocacy work because of the time sensitivity of the subject matter. The importance of trust and good working relationships in advocacy is an interesting topic in the context of an organization like Greenpeace.

Harkki acknowledged that in the early 2000s, Greenpeace and northern Sámi of Finland were effectively strangers to one another. In the beginning, "we didn't know each other. I think trust mostly exists between individuals and when we started working in the Sámi homeland, none of the Greenpeace staff members knew almost anyone there. Of course we had to start from scratch".[19] Over time, it appears that trust grew, at least in terms of willingness to work together on the forestry issue for which both sides had overlapping concerns.

Harkki elaborated, in response to the question, are people in Lapland more comfortable now being open about working with Greenpeace after 15 years of working together, to which she answered that essentially, yes, they are[20]

I think people are now comfortable talking about it openly. And the media is now more neutral to us nowadays and I think that makes it easier, that people don't need to feel that they are alone if they speak up or cooperate with Greenpeace. So yes, I do think it's become easier but I also think it's the young Sámi generation that's very different. [*Interviewer: How so?*] Some good things have happened since the 1990s. The local schools in the area are teaching Sámi again. There

> were some decades where this was not done. And young people seem to be very proud of their cultural heritage and the Sámi culture and I think it feels like a natural part to them to defend it. I think that has changed the whole atmosphere from what it was 15 years ago where also the forest industry was more present in the area, where the oppression on the reindeer herders was really quite bad. They actually had to be quite afraid and received death threats when they spoke in media. So it was really like only the bravest who didn't have anything to lose wanted to stand up and be the faces of the campaign. Everything has shifted now. It's more natural, easier.[21]

Time to get to know one another seemed to be an important ingredient in the evolution of the Greenpeace–Sámi relationship in Finland. Furthermore, acknowledgement of the experiences of violence also helped to foster the relationship.

What is fascinating about Harkki's and Christensen's reflections on Greenpeace's work with Sámi, particularly working with local peoples directly involved in and affected by the forestry debate, is how Greenpeace's campaigning with Sámi in Finland was in many respects the complete opposite of what the organization did during the anti-sealing movement in the 1970s. As previously discussed in Chapter 3, when Greenpeace first got involved in the anti-sealing movement, it interjected itself into the local issue of seal conservation (Patey 1990; Allen 1979). It then sought out local ally-ship, but the relationship that Greenpeace pushed for with locals was not handled with care – neither the push to create the alliance nor to swiftly abandoning it –within a year for strategic reasons (Dale 1996, 91; also see Patey 1990; Harter 2004; Burke 2020, 2021).

Also, in contrast to the efforts taken in the 2000s engagement in Northern Finland, Greenpeace did not display sufficient care during its anti-sealing protests about the peoples – Indigenous and non-Indigenous – who would be impacted by the actions, attitudes, and behaviours they displayed and encouraged, as admitted in part by Greenpeace Canada in its 2014 apology (Kerr 2014). In doing so Greenpeace inflicted and contributed to violence and harm by promoting simplistic narratives about the economic and cultural importance of seal hunting and remained silent as some of its members, allies and supporters inflicted forms of physical, economic, psychological and cultural violence against vulnerable peoples (Royal Commission on Seals and the Sealing Industry in Canada 1986; Woods 1986; Sinclair et al. 1989; Burke 2020, 2023). It also did not do enough to acknowledge or stop harms to Inuit and other Indigenous peoples (e.g. Mi'kmaq and Innu peoples of Newfoundland and Labrador) who became unintended secondary targets of anti-sealing activism (Kerr 2014; Burke 2020, 2021, 2023).

The reported effort given to cultural sensitivity and awareness while working with Sámi on forest conservation in the 2000s reflects a positive shift by some within Greenpeace to engage on northern issues by doing things differently as it pertains to relations with local peoples. Greenpeace chose not to replicate the past problematic dynamics of its campaigning, and that of other environmentalists, including viewing "humans as a 'cancer'" and using its "stigmatizing discursive power to label opponents as irrational and harmful" (Nyyssönen 2022, 5). It appears that in the case of northern Finnish old-growth forest protection, Greenpeace tried – success fully for its agenda and the interests of some concerned local Indigenous stakeholders and rightsholders – to take a different approach.

Conclusion

Overall, the Greenpeace–Sámi of Finland reindeer herder alliance in the early 2000s pushed for the preservation of grazing areas for reindeer. The overarching approach to the forestry issue in Finland, starting with Greenpeace only getting involved after being approached by concerned local Sámi and then working to learn about locals. Greenpeace members on the ground showed restraint by not trying to take over the campaign, and focusing on helping to elevate local voices in the areas of shared concern. This approach is a testament to the power of listening and the potential benefits of trying something new. In stark contrast to the complex legacy of Greenpeace's involvement in the anti-sealing movement in the 1970s–1980s, the relationship with the Sámi in Finland worked in terms of achieving satisfactory outcomes for all parties of the alliance. It also does not appear to have damaged the potential for future productive alliances in other areas of possible overlapping concern. But what was done with Sámi reindeer herders was unique for Greenpeace in the 2000s.

As the next two chapters illustrate, when Greenpeace activists show a level of cultural respect, demonstrate restraint in their campaigning approach, and listen to local stakeholders and rightsholders, the organization can have a positive effect. However, the ability of an organization of the size of Greenpeace to navigate the nuances of campaigning legacies and learn lessons, and implement and remain committed to change on a large scale, can be incredibly difficult. It is dependent on the quality and strength of leadership and campaigners both in the field and working centrally to: define Greenpeace's stances on issues; set the agenda; coordinate positions, messaging, and troubleshoot discontent expressed by supporters, donors, and different offices and allies; and manage media expectations and coverage.

Notes

1 Other common spellings: Saami and Sami.
2 Sápmi encompasses large segments of northern Norway, Sweden, and Finland, and part of neighbouring Russia on the Kola Peninsula.

Much of this area in the Nordic countries is commonly known in English as Lapland.

3 Part of the internal challenge Sámi face with regard to reindeer herding and discussing economic opportunities to aid self-determination is rights protection. The reindeer herding dimension of traditional Sámi life and culture received a lot of legislative focus within the Nordic states, such as Sweden, while other functions and use of traditional Sámi territories by Sámi peoples are less protected (Mörkenstam 2005; Wheelersburg 2008; Cederberg 2010).

4 In this instance, the Nordic countries referred to are Norway, Sweden, and Finland. Sápmi does not extend into the other Nordic countries of Denmark (including Kalaallit Nunaat/Greenland and the Faroe Islands) or Iceland.

5 According to Vuojala-Magga et al. (2011), as of 2011

> Finland has 56 herding cooperatives or profit-making herding units, each differing in size as defined by surface area and number of reindeer. Each herder in a cooperative has personal rights and obligations according to the number of reindeer he or she owns.

Of these 56 cooperatives, 8 are Sámi-based in Finnish Sápmi (Vuojala-Magga et al. 2011, 228). However, according to the website of the Finnish Reindeer Herders' Association, there are "54 different cooperatives" (Reindeer Herders' Association n.d.), but there is no date on this information on the official website.

6 In recognition of long-standing issues of discrimination against Sámi in Finland, "the Finnish government established [a] Truth and Reconciliation Commission to address the historical treatment of the indigenous Sami population and promote the attainment of the Sami people's rights" on October 28, 2021 (Library of Congress 2021).

7 Disputes about commercial enterprise between Sámi reindeer herders and the state forestry in Finland date back to the 1950s, with the dispute extending beyond Sámi reindeer herders to include ethnic Finnish herders, too (Lawrence and Raitio 2006, 1).

8 Reindeer herding in Finland has been organized in formal, state-imposed cooperatives since 1898, though "herding was organized as far back as the 1700s" (Vuojala-Magga et al. 2011, 228). In contrast with Norway and Sweden, "[i]n Finland, both Finns and Sami have similar rights to own reindeer and practice herding, whereas in Sweden and Norway, only the Sami have the right to practice the livelihood" (Vuojala-Magga et al. 2011, 228).

9 Since the campaigning in the early to mid-2000s, Metsähallitus has made more of a concerted effort to acknowledge reindeer husbandry and the need for communication, cooperation, and agreement between interested parties for the coexistence of industries and practices in a shared space, with its primary negotiating partners being the Reindeer Herders' Association and the various cooperatives that it represents (Metsähallitus n.d.).

10 The migratory pattern in Finland and some parts of Sweden is in contrast to

> [R]eindeer in Norway and Sweden and on the Kola Peninsula in northwestern Russia [which] usually spend the summer on the tundra or in coastal areas foraging on green vegetation. For the winter, they are moved to boreal forests or taiga, where they feed on terricolous lichens.
>
> (Vuojala-Magga et al. 2011, 228)

11 Greenpeace was an ally with Sámi reindeer herders (via cooperatives) and the Sámi Council;

> The Sami Home Region (Sapmi) in northernmost Finland has eight Sami herding cooperatives. Reindeer herding in each of these cooperatives is based on Sami customs and practices and organized in keeping with the siida system. A siida (in different Sami languages: cearru, siida, sita, kite) is traditionally defined as an extended family or kin group consisting of economically independent households.
>
> (Vuojala-Magga et al. 2011, 228)

12 Sini Harkki was an active member of Greenpeace Nordic at the time of the interview with her on September 21, 2018.

13 Interview by Danita Catherine Burke with Sini Harkki of Greenpeace Nordic, September 21, 2018.

14 Interview by Danita Catherine Burke with Sini Harkki of Greenpeace Nordic, September 21, 2018.

15 Mads Flarup Christensen was the Executive Director of Greenpeace Nordic at the time of his interview in January 2019, based in Copenhagen, Denmark.

16 Interview by Danita Catherine Burke with Mads Flarup Christensen of Greenpeace Nordic, February 7, 2019.

17 In Finland, the "rights and interests of the cooperatives are protected by the Reindeer Herders' Association, which is administered by the Ministry of Agriculture and Forestry" (Vuojala-Magga et al. 2011, 228).

18 Interview by Danita Catherine Burke with Sini Harkki of Greenpeace Nordic, September 21, 2018.

19 Interview by Danita Catherine Burke with Sini Harkki of Greenpeace Nordic, September 21, 2018.

20 Harkki also commented in response to a question about how the cooperation between Greenpeace and Sámi reindeer herders is being received by the Sámi community more generally, that

> I think very well. When we started working together 15 years ago talk and also media trying to create an image that we are using the Sámi for our campaigning purposes, which was very difficult for the people themselves because the reindeer herders that were involved were protecting their livelihoods in a really difficult situation to start with. They got this pressure. [where do you think the pressure

was coming from?] I think it was mainly coming from forest company that was doing the logging, but then they also had some employees that were Sámi themselves and the local media is, unfortunately, very hostile to Sámi rights in general. They [the local media] were writing a lot about it and you didn't see any really neutral stories. That was then. Now it's very different. We had very neutral media and one of our activities that we did together after the course had something like 130 local participants and it was close to a town in the Sámi area that has something like 1000 inhabitants so it was a pretty big turnout. The Sámi standing up for their rights proudly has definitely become more accepted locally.

21 Interview by Danita Catherine Burke with Sini Harkki of Greenpeace Nordic, September 21, 2018.

References

Allen, J. 1979. "Anti-Sealing as an Industry." *Journal of Political Economy* 87(2): 423–428.

Borchert, Nanna. 2001. "Land Is Life: Traditional Sámi Reindeer Grazing Threatened in Northern Sweden." *Svenska Sdddamernas Riksförbund*. http://www.oloft.com/landislife.pdf

Broderstad, Else Grete. 2022. "International Law, State Compliance and Wind Power: Gaelpie (Kalvvatnan) and Beyond." In *Indigenous Peoples, Natural Resources and Governance: Agencies and Interactions*, edited by Monica Tennberg, Else Grete Broderstad and Hans-Kristian Hernes. Routledge, 16–38.

Burke, Danita Catherine. 2020. "Re-establishing legitimacy after stigmatization: Greenpeace in the North American North." *Polar Record* 56 (e26): 1–12.

Burke, Danita Catherine. 2021. "The Case for a Greenpeace Apology to Newfoundland and Labrador." *The Northern Review* 51(1): 173–187.

Burke, Danita Catherine. 2023. *Cultural Violence, Stigma and the Legacy of the Anti-Sealing Movement*. Routledge.

Cederberg, Charlotta. 2010. "The Balance between the Right to Reindeer Herding and other Interests of Land Use in Sápmi: A Labour and Human Rights Perspective on the Situation of the Swedish Sami." *Faculty of Law, University of Lund*. https://lup.lub.lu.se/student-papers/search/publication/1690998

Cepinskyte, Agne. 2018. "The Arctic Railway and the Sámi: Reconciling National Interests with Indigenous Rights." *Finnish Institute of International Affairs*. FIIA Briefing Paper, December 2018/254. https://fiia.fi/en/publication/the-arctic-railway-and-the-sami

Cultural Survival. 2015. "Saami vs. Metsähallitus: The Case for Corporate Recognition of Indigenous Rights." February 25. https://www.culturalsurvival.org/publications/cultural-survival-quarterly/saami-vs-metsahallitus-case-corporate-recognition

Dale, S. 1996. *McLuhan's Children: The Greenpeace Message and the Media*. Between the Lines.

Dubash, Navroz K. and Michael Oppenheimer. 1993. "Modifying the Mandate of Existing Institutions: NGOs." In *Confronting Climate Change: Risks, Implications and Responses*, edited by Irving M. Mintzer. Stockholm Environment Institute. Cambridge University Press.

Greenpeace USA. 2010. "IKEA Announces New Policy Commitment on Ancient Forests." July 6. https://www.greenpeace.org/usa/news/ikea-announces-new-policy-comm/

Harter, John-Henry. 2004 "Environmental Justice for Whom? Class, New Social Movements, and the Environment: A Case Study of Greenpeace Canada, 1971–2000." *Labour/Le Travail* 54(2004): 83–119.

Howlett, Catherine and Rebecca Lawrence. 2022. "Emerging Governance Mechanisms in Norway: A Cautionary Note from the Antipodes." In *Indigenous Peoples, Natural Resources and Governance: Agencies and Interactions*, edited by Monica Tennberg, Else Grete Broderstad and Hans-Kristian Hernes. Routledge.

Kerr, Joanna. 2014. "Greenpeace Apology to Inuit for Impacts of Seal Campaign." *Greenpeace*, June 24. https://www.greenpeace.org/canada/en/story/5473/greenpeace-apology-to-inuit-for-impacts-of-seal-campaign/

Kuokkanen, Rauna. 2024. "The Problem of Culturalizing Indigenous Self-Determination: Sámi Cultural Autonomy in Finland." *The Polar Journal* 14(1): 148–166.

Lawrence, Rebecca and Kaisa Raitio. 2006. "Forestry Conflicts in Finnish Sápmi: Local, National and Global Links." *International Work Group for Indigenous Affairs (IWGIA)*, March 21. https://www.iwgia.org/images/publications/IA_4-06_Finland.pdf

Library of Congress. 2021. "Finland: Sami Truth and Reconciliation Commission Established." https://www.loc.gov/item/global-legal-monitor/2021-11-18/finland-sami-truth-and-reconciliation-commission-established/

Metsähallitus. n.d. "Northern and Eastern Finland Are Special Areas for Metsähallitus." https://www.metsa.fi/en/lands-and-waters/planning-of-state-owned-areas/special-regional-features/

Mörkenstam, U. 2005. "Indigenous Peoples and the Right to Self-Determination: The Case of the Swedish Sami People." *The Canadian Journal of Native Studies* XXV(2): 433–461.

Ministry of Petroleum and Energy. 2023. "Norwegian Government Apologises to Sámi Reindeer Herders on the Fosen Peninsula." *Government of Norway*, April 3. https://www.regjeringen.no/en/aktuelt/norwegian-government-apologises-to-sami-reindeer-herders-on-the-fosen-peninsula/id2965357/

News.cision 2010. "Campaign for Northern Forests by Indigenous Saami and Greenpeace Ended Successfully in Finland." https://news.cision.com/, December 10. https://news.cision.com/greenpeace/r/campaign-for-northern-forests-by-indigenous-saami-and-greenpeace-ended-successfully-in-finland,c532196

Nilsen, Thomas. 2021. "Lapland Regional Council Rejects Arctic Railway." *Arctic Today*, May 17. https://www.arctictoday.com/lapland-regional-council-rejects-arctic-railway/

Nyhlén, Sara, Katarina Giritli Nygren, Anna Olofsson and Johanna Bergström. 2018. "Human Security, Risk and Sustainability in the Swedish Policy for the Arctic." In *Human and Societal Security in the Circumpolar Arctic: Local and Indigenous Communities*, edited by Kamrul Hossain, José Miguel Roncero Martín, and Anna Petrétei. Brill Nijhoff.

Nyyssönen, Jukka. 2022. "Frame Alignment between Environmentalists and the Sámi in the Forest Dispute in Inari, Finland Until the 2000s—Competing Conservation Needs and Obstacles for Co-Living With the Non-Human." *Frontiers in Conservation Science* 3: 925713. doi:10.3389/fcosc.2022.925713

Patey, Francis. 1990. *A Battle Lost: An Unsuccessful Attempt to Save the Seal Hunt*. Robinson-Blackmore Printing and Publishing Ltd.

Quinn, Eilís. 2019. "The Arctic Railway." *CBC: Eye on the Arctic*, September 22. https://newsinteractives.cbc.ca/longform/the-arctic-railway/

Reindeer Herders' Association. n.d. "Reindeer Herding Cooperatives." https://paliskunnat.fi/reindeer/reindeer-herding/cooperatives/

Royal Commission on Seals and the Sealing Industry in Canada. 1986. "Volume 1: Economic Options, Prospects and Issues." Memorial University of Newfoundland, Queen Elizabeth 2nd Library, Centre for Newfoundland Studies Stacks SH362.D761986.

Sinclair, Peter, Robert Hill, Cynthia Lamson and H.A. Williamson. 1989. "Social and Cultural Aspects of Sealing in Atlantic Canada." Report for the Royal Commission on Seals and the Sealing Industry in Canada. Institute of Social and Economic Research, Memorial University of Newfoundland.

Vuojala-Magga, T., M. Turunen, T Ryyppö and M. Tennberg 2011. "Resonance Strategies of Sámi Reindeer Herders in Northernmost Finland during Climatically Extreme Years." *Arctic* 64(2): 227–241.

Wheelersburg, Robert P. 2008. "The Need to Conduct Studies of Swedish Saami Reindeer-Herder Subsistence Behaviours: A Case of Indigenous Resource-Use Rights." *The Northern Review* 28: 161–180.

Woods, S.J. 1986. "The Wolf at the Door: The Anti-Harvest Campaign Strikes at the Heart of North Aboriginal Economies." *Northern Perspectives* 14(1): 1–8.

5 Seismic Testing and Inuit Rights
Kangiqtugaapik/Clyde River, Nunavut

When considering Greenpeace's work to overcome challenging dimensions of its campaigning legacy, arguably the clearest example of the direct benefits for tackling past mistakes head-on is Greenpeace's 2014 apology to Inuit, Indigenous, and coastal peoples in Canada (Kerr 2014). This apology is directly connected to Greenpeace's ability to work with Kangiqtugaapik/Clyde River, a small Inuit community situated along the northeastern coast of Baffin Island in Nunavut, Canada (Travel Nunavut n.d.). In 2014, the unassuming community of 850 people found itself at the centre of what would eventually become a landmark Supreme Court of Canada ruling on the duty to consult Indigenous communities. Greenpeace, unexpectedly, played a central role in making the community's defence of its Indigenous rights and assertion of its views on planned seismic testing possible.

This chapter unpacks Greenpeace's ally-ship with Kangiqtugaapik/Clyde River and the significance of Greenpeace's 2014 apology for elements of its anti-sealing legacy in helping it to start to mend relations with Inuit in Canada. Chapter 6 stresses, with personal reflections from Clyde River's mayor, Jerry Natanine, the role of individual activists in Greenpeace in helping to forge a revised approach to local engagement with renewed cultural sensitivity and relationship building to foster organizational growth and healing from the legacy of the anti-sealing movement.

Kangiqtugaapik/Clyde River, the National Energy Board, and Seismic Testing

In 2014, the community of Clyde River was informed that the National Energy Board (NEB) of Canada had approved a five-year seismic testing project proposed by a Norwegian consortium in the waters near the community. The consortium that planned to do the seismic testing consisted of Petroleum Geo-Services Inc., Multi Klient Invest AS, and TGS-Nopec Geophysical Company ASA (Greenpeace Canada 2017). Leading up to

DOI: 10.4324/9781003674023-5

the NEB decision, the community was made aware of the proposal, and there was some local consultation, but the community and its people were against the proposed plans (Supreme Court Judgments 2017; Tasker 2017).

The residents voiced their concerns and opposition during the consultation process and asked questions about the environmental and ecological implications of seismic testing. Seismic testing is a debated, though presently essential, method in marine oil and gas exploration. The problem with seismic testing is that "[h]uman-made noise is widely recognized as a pollutant and a key stressor that may have long-term detrimental effects on marine ecosystems" (Affatati and Camerlenghi 2023, 1). Leading global energy and petrochemical company Shell[1] refers to seismic testing as marine seismic surveys and states that they believe that

> [M]arine seismic surveys are a safe mapping technique for gathering information about the location and properties of oil or gas that may be present deep below the seabed of a given area during the exploration and production phases of a project.
>
> (Shell n.d.-a, Marine seismic surveys)

However, Affatati and Camerlenghi (2023, 2) in their work on the effects of marine seismic surveys state: "An increasing number of scientific studies suggest that the impulses generated during seismic surveys adversely impact marine animal species, eliciting masking, behavioral responses, and changes in acoustic repertoires".

The people of Kangiqtugaapik/Clyde River had previous experience with seismic testing and did not want it to occur near their community again. Mayor Natanine, who led Kangiqtugaapik's/Clyde River's NEB appeal and Supreme Court challenge, spoke of his father's recollections of the negative impact of seismic testing with dynamite that had occurred near the community in the 1970s. Natanine recalled his father "said that seals became deaf, their ears were bleeding and puss coming out when they'd catch them" (Burke 2021, 107). Natanine and community members were gravely concerned that seismic activity would destroy the local ecosystem, given its documented impacts on marine life and their previous experience with such testing.

From the community members' perspective, their questions, concerns, and ultimate disapproval of the proposed testing were sidelined by the NEB, which approved the consortium testing licence despite overwhelming local opposition.[2] Greenpeace became involved with Kangiqtugaapik/Clyde River after the initial NEB decision. This ally-ship resulted in the community appealing the NEB decision up to the Supreme Court of Canada.

The Supreme Court of Canada ruled in 2017 that the NEB approved the licensing despite inadequate community consultation processes with Inuit rightsholders in the area (Supreme Court Judgments 2017). The Supreme Court concluded that "the consultation and accommodation efforts in this case were inadequate and fell short in several respects... Therefore, the Crown breached its duty to consult in respect of the proposed testing" (Supreme Court Judgments 2017; also see Tasker 2017).[3] This ruling halted the testing plans to the relief of the local residents.

Working with Greenpeace and the Supreme Court Challenge

Greenpeace and Clyde River came together in the space between the initial NEB decision and the community's efforts to mount an appeal. According to the rules around the authorization of proposals, there was a 30-day window during which communities could file an appeal to the Federal Court of Appeal in Canada as a last effort to stop or delay whatever development project the NEB had approved (Nunatsiaq News 2014).[4] Realistically, however, 30 days is a very short period of time, and most small communities cannot afford the costs associated with an appeal, which puts a lot of pressure on communities to stop fighting (Burke 2021).

Jessica Wilson of Greenpeace Canada[5] was one of the frontline members of Greenpeace working directly with Clyde River Mayor Jerry Natanine and his community.[6] In Wilson's opinion, the consultation process conducted was completely inadequate.

> First of all, the company that was going to do this blasting applied for this permit and received it despite an absolutely abhorrent, dehumanizing consultation process that barely paid lip service to the rights of Inuit and Inuit but also as human beings. They weren't able to answer any of the questions that would be more important to this community: What are the impacts on the marine mammals? What are the impacts on everything from the Arctic char to the seal to beluga to narwhal to blue whales to polar bears? These are all animals for whom their ears are their everything...This consortium of Norwegian companies that had received this permit, this permit allowed them to conduct this blasting for the entire ice free season, essentially May to September, 24 hours a day. These blasts would be going off every 10-5 seconds, 24 hours a day, repeatedly and they had the gall to say that there is no concrete evidence that this has any impact on marine mammals.[7]

At the time of the initial NEB decision in 2014, Kangiqtugaapik/Clyde River and Mayor Natanine found themselves with little time to appeal and few financial resources to proceed. Natanine reached out to many

NGOs like the World Wide Fund for Nature and local Inuit governments and corporations for aid, but nobody thought the appeal had a chance of success, and they were therefore unwilling to commit resources to the community's appeal (Burke 2021).

It is here, in this space of the 30-day appeal deadline, that Greenpeace Canada enters the picture. Importantly, in the context of this chapter, it is how Greenpeace Canada becomes involved and how it conducts itself during its involvement that is most significant for learning about its efforts to change Greenpeace's northern/Arctic advocacy approach. Specifically, Greenpeace Canada's involvement with Clyde River reveals two important features of how cooperation between the NGO and the Inuit community was possible given the legacy of the anti-sealing movement which undermined Greenpeace–Inuit relations in Canada (Phelps Bondaroff and Burke 2014; Burke 2023, 2020): (1) the power of an apology, in this instance an apology in 2014 from Greenpeace Canada to Inuit of Canada, Indigenous and coastal peoples for their role perpetuating colonial attitudes, actions and behaviours during its the anti-sealing campaigning (Kerr 2014)[8] and (2) the necessity of local instigation and leadership in building an alliance and running the campaign (Burke 2021).

According to Wilson, Greenpeace Canada's alliance with Kangiqtugaapik/Clyde River started as a result of the apology Greenpeace Canada had publicly issued for its anti-sealing legacy; the apology incidentally came out around the same time as the community faced its 30-day appeal countdown timeline (also see Kerr 2014). As previously stated in Chapter 4, the apology included an acknowledgement that

> We have a responsibility — not just as an organization that once campaigned against the commercial hunt, but also as conscious, socially responsible human beings — to right wrongs, to actively stop the spread of misinformation, and to decolonize our thinking, our language and our approach.
>
> (Kerr 2014)

The apology started a chain of events for Greenpeace that were unanticipated, as it was not specifically targeted towards Kangiqtugaapik/Clyde River. The apology was also not made with any expectation of reply from an Inuit community, organization, or government. Wilson noted that

> It all started with the publication of the organisational apology to the Inuit, which was published on June 21st, 2014, I think, and this was a step that we, the leadership of Greenpeace Canada and myself as the head of the Arctic campaign, knew was absolutely [a] fundamental first step. Greenpeace Canada hadn't operated in the Arctic in decades.[9]

The importance of the Greenpeace apology cannot be overstated. It set the scene for eventual cooperation between the organization and Kangiqtugaapik/Clyde River.[10] Natanine reflected that "A week or two before [the appeal filing deadline] Greenpeace had written an apology for the effect that the seal ban had on Inuit culture and our way of live, how it was effected" (Burke 2021, 108). However, it must be noted that Greenpeace was not Natanine's first choice of an ally.[11]

There is profound dislike towards Greenpeace amongst Inuit of Northern Canada because of Greenpeace's leading role in defining and normalizing anti-sealing activism and the impact that anti-sealing campaigning had, and continues to have, in undermining Inuit economies and rights. For their part, Greenpeace members are aware of this. As Jessica Wilson stated, "I don't think we can understate the depth of the hatred and resentment between Inuit and Greenpeace".[12] Natanine echoed, from a personal perspective, a long-held, deep-seated hatred for Greenpeace; "I hated Greenpeace all my life. You know, there was an enemy and it was Greenpeace" (Burke 2021, 107).

When Natanine first started looking for help to appeal the NEB decision, Greenpeace was not on his radar of potential allies, but after contacting

> Oceans North, Eco-Justice, and a couple of others. We contacted our Inuit organizations—NTI [Nunavut Tunngavik Incorporated] and QIA [Qikiqtani Inuit Association] and someone contacted our Member of Parliament at the time, but we couldn't find anyone [who could or would help] and the deadline was coming.
>
> (Burke 2021, 107)

It was only once other opinions were exhausted that Natanine decided to turn to Greenpeace with a request for help, with only three days before the appeal deadline, and he was only willing to consider the option because it was preceded by Greenpeace's public apology for its anti-sealing campaigning and its impact on the Inuit of Canada (Burke 2021, 107).

Natanine reflected that Greenpeace's openness to his request for help reflects well on the organization's willingness to take a chance to build on its public apology and step up to try and help people it had previously harmed. Greenpeace stepped up when other established environmental organizations that were better received locally were unwilling to help and use their limited resources in a situation that, at times, appeared hopeless.

> At the time it didn't look like we'd win. I didn't know this when it was happening, but afterward I started hearing things about why and it was that nobody thought we could win. That was the only reason....We contacted WWF [World Wide Fund for Nature/World Wildlife Fund] in Iqaluit and Oceans North who also have an office in Iqaluit. They had

> the reason of funding for why they couldn't help. After all that, I was talking with Greenpeace, after we first started after talking with all these organizations and Inuit organizations for help and no one could help.
>
> (Burke 2021, 108)

The 2014 apology created a window of opportunity for dialogue.[13] But the relationship was an alliance of convenience, as Inuit trust in Greenpeace is minimal due to the anti-sealing legacy and the decades of hardship that the fallout from the anti-sealing movement caused for Inuit peoples and societies (Woods 1986), as well as its broad approach against non-renewable resource development (Burke 2020).

Natanine noted that the alliance was a last hope for Kangiqtugaapik/ Clyde River; in

> the beginning we were being told [by other actors] that there just isn't any money and when it was over and we seen how much it cost, we thought how could there not be any money?! Basically, that's why we reached out to Greenpeace.
>
> (Burke 2021, 108)

Greenpeace provided a last-minute chance for the community, which, according to Wilson, was the following:

> The feeling from this community was very much that they were fighting for their life. So Jerry had a contact with Ecojustice and he reached out to them and Ecojustice put Jerry and I in touch…At this point I think there were 72 hours left before the deadline to file [an appeal] was there…I can reach out to every lawyer I know to fight this and we are happy to fund this and see where it goes. Is that okay with you? And he said yes. Let's go for it.[14]

For the community, Greenpeace provided access to resources that were needed on short notice to file an appeal.

Natanine was very forthright about what Greenpeace brought to the table and what the community wanted and needed during the alliance.

> There was the money and other things, like the network they have with the membership they have and it really worked out well for media purposes. They handled all our media, all our lawyer or money, but it was a really good relationship because before they put out anything in regard to our case and in regard to our relationship, we always had the last word here in Clyde River; if we liked it or not. That was really great for us because we were not only changing as a community, getting over our hatred and working with an organization we had regarded as our

> enemy, but also on their side too. They had never worked with an Inuit group of any kind and they know that they are looked at in a certain way. So on both sides there was all this change happening.
>
> (Burke 2021, 108)

The corporation appears to have done more than just address the immediate needs of the community; it set the stage for Greenpeace to demonstrate that it was willing to be somewhat different as an organization from how it presented itself and treated others during its anti-sealing campaigning years.

As such, for some in Kangiqtugaapik/Clyde River, such as Natanine, the Kangiqtugaapik Inuit–Greenpeace relationship evolved somewhat:

> In the beginning, because we are strangers and at the time it seemed like it was going to stay like that and that we were going to be far apart from each other; never see each other, just handle the money, lawyer and media attention that we need for all of our support and I'd never have to meet any of them.
>
> (Burke 2021, 108)

Over time, however, "we made great friends. I think life lasting friendships were made. We all treated each other really nicely" (Burke 2021, 108). Importantly for the relationship evolution, the apology to the Inuit was never intended specifically to appeal to Kangiqtugaapik/Clyde River or to explicitly elicit an acceptance or action by the Inuit in reply. As Wilson recalls, "We didn't know what to expect with this apology, we just knew it was the right thing to do and a very clear first step".[15]

Arguably, a central reason why the working relationship between Kangiqtugaapik/Clyde River and Greenpeace resulted in positive views from Natanine towards the Greenpeace members he worked with is how the working relationship was constructed. As Wilson reflected on her time with Natanine and the community,

> Jerry [Natanine] and the community of Clyde River, they are the ones on the front line of this fight. They know what's going on, they know what's at stake. We just asked them to tell their stories and we helped share them with our global audiences. I think the role that we played in this instance was that we did have a big global base to draw upon, a bigger megaphone that we could kind of pass to them. It was an organizational exercise in passing the mic, so all of their quotes, all of their social media, all of the narrative and communications came directly from Clyde. We had a very respectful sign off process. We didn't take a single move that wasn't led by, approved by and designed by the community. We didn't want to put any spin on anything. We just wanted to

> take their words and their stories and try to share them with a bigger audience than they might otherwise be able to access. And I think that's the role of big NGOs in these spaces. I think we might be seeing a shift globally in different parts of the world in better ally-ship, deeper ally-ship. NGOs are often accused of parachuting in and out of issues, of being the 600 pound gorilla that storms into these spaces. That's not how we wanted to work in this case. I think there are really beautiful examples of not working in that way, of working in respectful ally-ship where it is really community led, community driven. We are there to help lift up and support rather than take over. I think the entire team we were working with at Greenpeace Canada, it was a really really beautiful team of people and that was just fundamental for all of us. I think we were really aligned. We were aware of what a privilege it was to be given kind of a shot by Clyde River to show up and prove that we were a different organisation and that we wanted to be different and that we were not taking our marching orders from Greenpeace campaigns of the past but rather from the community of the moment.[16]

Natanine also acknowledged the importance of the community's voice in final decision-making in the alliance.In Natanine's view,

> any time Greenpeace was going to do any media work or reach or fundraising, all these things, you have to agree up front. One thing I said was very beneficial to us is that we have to have the last say before anything goes out.
>
> (Burke 2021, 109)

Given the trend of marginalizing Indigenous voices of environmental activists, as discussed in Chapters 2 and 3, the pattern of working, as noted by Natanine, is a departure from this problematic approach to conservationism in the environmental movement. Interestingly, Wilson made a point to emphasize that Greenpeace – Greenpeace Canada and Greenpeace International – is not pointing to the Kangiqtugaapik/Clyde River ally-ship as absolution for Greenpeace's anti-sealing campaigning past. Wilson reflected that

> We were always very careful to not lean on the example of our work with Clyde River as a way to absolve ourselves from past sins. … It wasn't like that at all and we were very careful to not lean on Clyde River like that or use them to prop up this vision how we wanted the organisation to be perceived.[17]

Thus far, the Kangiqtugaapik/Clyde River alliance stands a bit of a one-off for a campaign of its scale and level of success, as Sune Scheller of

Greenpeace Nordic observed: "I think what's been going on in Clyde River is unparalleled elsewhere in the Arctic. That is the amount of resources and priority that has been given is more than what's happened in Greenland".[18] Recognizing that one good campaign with one community does not resolve all past issues with sealing is a good point to recognize. Embodying the purported contrition for the past and working to not replicate its patterns in current and future work will require self-discipline for the organization and its members.

In Kalaallit Nunaat/Greenland, for example, Greenpeace's work is more low-key, such as "visiting at schools and tours along the coast to the settlements", but such work is also important as part of an overarching strategy to engage with Arctic peoples.[19] Greenpeace is mistrusted by many Inuit audiences in Kalaallit Nunaat/Greenland, as well as Canada, due largely to its anti-sealing legacy (Wessendorf 2011, 27). Greenpeace is working to build inroads and public dialogue with local governments, rightsholders, and stakeholders in Kalaallit Nunaat/Greenland (Burke 2020), but it has been challenging as its efforts to balance public criticism for its anti-sealing campaigning have faced pushback from within Greenpeace's international network of offices, supporters, and donors.

This challenge was most noticeable in 2011 when a Greenpeace Nordic (Denmark) representative, Jon Burgwald, wore a seal skin vest in a local media interview in Kalaallit Nunaat/Greenland and expressed public support for traditional Kalaallit/Greenlandic sealing. Burgwald's actions were received well in Kalaallit Nunaat/Greenland, but the media story caught the attention of a Dutch journalist, and it became a big story in Europe that a Greenpeace member was wearing seal fur. The news story caused donors, supporters, and some organizational members to reportedly become concerned that Greenpeace now appeared to be supporting the killing of seals. The internal debate on Greenpeace's stance on seal hunting that resulted fed into the perception of organizational inconsistencies in its reconciliation of its anti-sealing legacy by Kalaallit/Greenlandic audiences and set back Greenpeace Nordic's efforts to re-establish trust with local audiences (Burke 2020).[20]

Conclusion

Greenpeace Canada's successful alliance with Kangiqtugaapik/Clyde River proved to be very beneficial for the community. While they lost their appeal to the Federal Court of Appeal (Zerehim 2016), they successfully appealed that ruling up to the Supreme Court of Canada, where it was determined that the consultation with the community was inadequately conducted and the licence issued by the NEB could not stand as a result (Supreme Court Judgments 2017).[21] The whole series of events that led to the ally-ship, however, hung on the coincidental timing of Greenpeace's apology in 2014. Without the 2014 apology, Greenpeace would not have been on the radar for Natanine as a potential source for help, and it would have been much harder for Natanine to personally consider, or to convince others in his community to consider, Greenpeace as an ally if the apology had not happened when it did. The apology created a window of opportunity, which Natanine explored.

Greenpeace's willingness to restrain itself and be open to the community's concerns and leadership in the joint campaign also proved essential to both the legitimacy of Greenpeace's involvement with the Inuit of Kangiqtugaapik/Clyde River and to the community's overall positive impression of Greenpeace as a current and future ally. It is impossible to generalize about the status of the relationship between Inuit and Greenpeace based on the Kangiqtugaapik/Clyde River ally-ship experience and Greenpeace's conduct during that specific campaign. However, the fact that the campaign was possible, perceived as positive locally and yielded successful results based on the community needs while in alignment with Greenpeace's agenda does indicate that when invited, motivated and culturally considerate, Greenpeace can engage in environmental issues that impact rural areas and peoples in ways that avert inflicting violence on locals while still leading to their desired outcomes and remaining consistent with the organization's core beliefs like in nonviolence, taking responsibility for its actions and working to find solutions to problems.

The events in Kalaallit Nunaat/Greenland in 2011 illustrate, however, that organizational priorities and tensions over to what extent Greenpeace can and will apologize for all aspects of the harms from anti-sealing activism are a challenge. Greenpeace's work to reconcile negative elements of campaigning history in the North and remain consistent is hard for an organization of its size. Greenpeace Canada has shown an effort to start the process of acknowledging of anti-sealing campaign legacy and that of its allies and contemporaries at the time. This contributed to the formation of the alliance with Kangiqtugaapik/Clyde River, but efforts in Kalaallit Nunaat/Greenland have been more froth. At this stage, it is too soon to determine how consistent Greenpeace will be with regard to its stance on seals, sealing, and Inuit self-determination and the use of seals as part of an economic and cultural path in that direction.

That said, the success with Kangiqtugaapik/Clyde River and the 2014 apology are positive signs of a least some internal shifting about the anti-sealing legacy and the benefits in terms of relationship and opportunity building that might be possible for Greenpeace if it more directly addresses this past. However, the pushback within Greenpeace's base support from a member of Greenpeace who was trying to exhibit cultural nuance about the killing and use of seals in Kalaallit Nunaat/Greenland in the 2010–2011 period and address the organization's anti-sealing legacy more fully, will act as a deterrent to others wishing to navigate and reconcile this aspect of its history with more depth going forward.

Notes

1 Shell's website technically states that it is "a global group of energy and petrochemical companies with more than 103000 employees in more than 70 countries" (Shell n.d.-b, Who we are).

2 Wilson recalled that the NEB:

> decided that sufficient consultation with the community had been done and that the permits stood and they [the consortium] could go ahead….If you see the documents, the transcripts of the consultations at Clyde River, dehumanizing is the word that comes to mind more often than not. It was a box ticking exercise for the National Energy Board. They didn't send any biologists up there, nobody who could actually speak to the things that would matter.
>
> (Interview by Danita Catherine Burke with Jessica Wilson of Greenpeace Canada, October 30, 2022)

3 In its judgment on the Clyde River (Hamlet) v. Petroleum Geo-Services Inc. case, the Supreme Court of Canada noted that in this instance,

> The NEB's approval process, in this case, triggered the duty to consult. Crown conduct which would trigger the duty to consult is not restricted to the exercise by or on behalf of the Crown of statutory powers or of the royal prerogative, nor is it limited to decisions that have an immediate impact on lands and resources. The NEB is not, strictly speaking, "the Crown" or an agent of the Crown. However, it acts on behalf of the Crown when making a final decision on a project application. In this context, the NEB is the vehicle through which the Crown acts. It therefore does not matter whether the final decision maker is Cabinet or the NEB. In either case, the decision constitutes Crown action that may trigger the duty to consult. The substance of the duty does not change when a regulatory agency holds final decision-making authority.
>
> (Supreme Court Judgments 2017)

4 Jerry Natanine stated that there were 31 days to appeal in his interview with Danita Catherine Burke (Burke, 2021), but the Nunatsiaq News (2014) states that the timeline was 30 days.

5 Jessica Wilson has worked for both Greenpeace Canada and Greenpeace International. During her time working on the Greenpeace alliance with Clyde River, they were part of Greenpeace Canada.

6 According to the Mayor of Clyde River during the seismic testing dispute, Jerry Natanine, "We worked with Jessica Wilson, Sara Chon, Joanna Curr. We mainly worked with them. There were others too, in the media department, but these three were really the leaders of Greenpeace up here at the time" (Burke 2021 109); Interview by Danita Catherine Burke with Jerry Natanine, Mayor of Kangiqtugaapik/Clyde River, Nunavut, Canada, December 19, 2018. The complete interview with Mayor Natanine is published in the *Arctic* journal's InfoNorth Series.

7 Interview by Danita Catherine Burke with Jessica Wilson of Greenpeace Canada, October 30, 2018.

8 Wilson stated in reference to the publication of the 2014 apology:

> We didn't know what to expect with this apology, we just knew it was the right thing to do and a very clear first step. On what was then called National Aboriginal Day and is now called National Indigenous Peoples Day in Canada, which is June 21st, we ran the apology in a couple of outlets. We ran it in Inuktitut in the North, in the Nunatsiaq news and we ran it in the south in iPolitics and published it to our [web]page. And we sent it to a number of Inuit organisations and just sort of said, we have no expectations from this but we want you to know that organisationally we feel that we made mistakes back then and we want to be accountable for it.
>
> (Interview by Danita Catherine Burke with Jessica Wilson of Greenpeace Canada, October 30, 2018)

9 Interview by Danita Catherine Burke with Jessica Wilson of Greenpeace Canada, October 30, 2018.

10 Wilson notes that there are those in Greenpeace leadership who are aware of just how detrimental Greenpeace's anti-sealing legacy is to the organization's relationship with Inuit throughout the Arctic:

> The Greenpeace of the day had this seal hunt campaign that for better or worse came to define us as an organisation for many, many years.... I think when you talk about the brand of an international organisation that can work for and against you. The burden of this big global brand is such that a lot of the campaigning work done by other organisations was sort of painted with the Greenpeace brush where the image of it for many people is that it was sort of all Greenpeace. But Greenpeace, the leadership of it at the time, had quite a diversion of opinions on the seal campaign. So once they had gone up to the North and met with some Inuit and began to understand the impacts of their campaigns, Bob Hunter and Paul Watson end up having a difference of opinion on this, led in a lot of ways to the creation of Sea Shepherd.
>
> (Interview by Danita Catherine Burke with Jessica Wilson of Greenpeace Canada, October 30, 2018)

11 Wilson also recalled that the time between the apology and the appeal deadline was very close:

> The apology landed a couple of weeks or so before the deadline for Clyde River to file an apology with the federal court of appeal to overturn a permit that had been approved for seismic blasting, which the industry innocuously calls seismic testing, in their waters.
>
> (Interview by Danita Catherine Burke with Jessica Wilson of Greenpeace Canada, October 30, 2018)

12 Interview by Danita Catherine Burke with Jessica Wilson of Greenpeace Canada, October 30, 2018.

13 Newell and Grant (2000, 240) state that

> A group's influence is often manifested by the acceptance of its ideas which capture the political imagination where there exists a "window of opportunity" for a new interpretation. If a concept or policy which NGOs are advocating resonates with an existing policy discourse, or institutional need, then it has a greater chance of being adopted.

14 Interview by Danita Catherine Burke with Jessica Wilson of Greenpeace Canada, October 30, 2018.
15 Interview by Danita Catherine Burke with Jessica Wilson of Greenpeace Canada, October 30, 2018.
16 Interview by Danita Catherine Burke with Jessica Wilson of Greenpeace Canada, October 30, 2018.
17 Interview by Danita Catherine Burke with Jessica Wilson of Greenpeace Canada, October 30, 2018.
18 Interview by Danita Catherine Burke with Sune Scheller of Greenpeace Nordic in Denmark, September 3, 2018.
19 Interview by Danita Catherine Burke with Sune Scheller of Greenpeace Nordic in Denmark, September 3, 2018.
20 For more on the fallout of Greenpeace in Greenland in 2010–2011 and the wearing of seal fur, see Chapter 2.
21 The Supreme Court of Canada judgment specifically stated with regard to the NEB consultation that

> While the Crown may rely on the NEB's process to fulfill its duty to consult, the consultation and accommodation efforts in this case were inadequate and fell short in several respects. First, the inquiry was misdirected. The consultative inquiry is not properly into environmental effects per se. Rather, it inquiries into the impact on the right itself. No consideration was given in the NEB's environmental assessment to the source of the Inuit's treaty rights, nor to the impact of the proposed testing on those rights. Second, although the Crown relies on the processes of the NEB as fulfilling its duty to consult, that was not made clear to the Inuit. Finally, and most importantly, the process provided by the NEB did not fulfill the Crown's duty to conduct the deep consultation that was required here. Limited opportunities for participation and consultation were made available. There were no oral hearings and there was no participant funding. While these procedural safeguards are not always necessary, their absence in this case significantly impaired the quality of consultation. As well, the proponents eventually responded to questions raised during the environmental assessment process in the form of a practically inaccessible document months after the questions were asked. There was no mutual understanding on the core issues — the potential impact on treaty rights, and possible accommodations. As well, the changes made to the project as a result of consultation were insignificant concessions in light of the potential impairment of the Inuit's treaty rights. Therefore, the Crown breached its duty to consult in respect of the proposed testing.
>
> (Supreme Court Judgments 2017)

References

Affatati, Alice and Angelo Camerlenghi. 2023. "Effects of Marine Seismic Surveys on Free-Ranging Fauna: A Systematic Literature Review." *Frontiers in Marine Science* 10(1222523): 1–21. https://doi.org/10.3389/fmars.2023.1222523

Burke, Danita Catherine. 2020. "Re-Establishing Legitimacy after Stigmatization: Greenpeace in the North American North." *Polar Record* 56(e26): 1–12.

Burke, Danita Catherine. 2021. "The Relationship between Kanngiqtugaapik/Clyde River and Greenpeace: An Interview with Mayor Jerry Natanine." *Arctic* 74(1): 106–110.

Burke, Danita Catherine. 2023. *Cultural Violence, Stigma and the Legacy of the Anti-Sealing Movement*. Routledge.

Greenpeace Canada. 2017. "Clyde River's case against seismic testing at the Supreme Court of Canada." July 26. https://www.greenpeace.org/canada/en/press-release/310/clyde-rivers-case-against-seismic-testing-at-the-supreme-court-of-canada/

Kerr, Joanna. 2014. "Greenpeace Apology to Inuit for Impacts of Seal Campaign." *Greenpeace*, June 24. https://www.greenpeace.org/canada/en/story/5473/greenpeace-apology-to-inuit-for-impacts-of-seal-campaign/

Newell, Peter and Wyn Grant. 2000. "Environmental NGOs and EU Environmental Law." In *Yearbook of European Environmental Law*, Edited by Han Somsen. Volume 1. Oxford University Press, 225–252.

Nunatsiaq News. 2014. "Clyde River Groups Challenge Nunavut Seismic Testing in Court." July 28. https://nunatsiaq.com/stories/article/65674clyde_river_groups_go_to_court_over_nunavut_seismic_testing/

Phelps Bondaroff, Teale N. and Danita Catherine Burke. 2014. "Bridging Troubled Waters: History as Political Opportunity Structure." *Journal of Civil Society* 10(2): 165–183.

Shell. n.d.-a "Marine Seismic Surveys." https://www.shell.com/what-we-do/oil-and-natural-gas/exploration/marine-seismic-surveys.html

Shell n.d.-b "Who We Are". https://www.shell.com/who-we-are.html

Supreme Court Judgments. 2017. "Clyde River (Hamlet) v. Petroleum Geo-Services Inc. Case number 36692." https://scc-csc.lexum.com/scc-csc/scc-csc/en/item/16743/index.do

Tasker, J.P. 2017. "Supreme Court Quashes Seismic Testing in Nunavut, but Gives Green Light to Enbridge Pipeline." *CBC News*, July 26. https://www.cbc.ca/news/politics/supreme-court-rulingindigenous-rights-1.4221698

Travel Nunavut. n.d. "Clyde River." https://travelnunavut.ca/regions-of-nunavut/communities/clyde-river/

Wessendorf, Kathrin. 2011. "The Indigenous World 2011." *The International Work Group for Indigenous Affairs*. https://www.iwgia.org/en/resources/publications/305-books/2977-the-indigenous-world-2011.html

Woods, S.J. 1986. "The Wolf at the Door: The Anti-Harvest Campaign Strikes at the Heart of North Aboriginal Economies." *Northern Perspectives* 14(1): 1–8.

Zerehim, Sima Sahar. 2016. "Clyde River Granted Leave to Appeal to Supreme Court on Seismic Testing." *CBC News*, March 10. https://www.cbc.ca/news/canada/north/clyde-river-granted-appeal-1.3484968

6 Protesting Oil Exploration and Working to Protect a Sacred Area

Arctic 30 and the Nenet and Khanty of Numto, Russia

Everything is different in Russia.[1] Perhaps this statement is a bit hyperbolic, but time and again, research and interviews with Greenpeace representatives (Burke 2022)[2] have noted that the operational conditions for environmentalists working in Russia are quite different from the conditions of colleagues working in other parts of the Arctic and North. Given Russia's relative political and media isolation during the Soviet era, Greenpeace's relationship with audiences in Russian society differs from the Western Arctic states. This does not mean that contemporary political actors, such as Indigenous organizations, are not aware of Greenpeace's campaigning legacy in broad terms, but they have not been directly impacted in the same way by Greenpeace's anti-sealing and anti-fur campaigning as other Indigenous actors, such as Inuit in Alaska, Kalaallit Nunaat/Greenland, and Canada were impacted. And to some extent, as is reflected in this chapter, some Indigenous actors in Russia remain more hesitant about open cooperation with Greenpeace, and are in solidarity with their brethren in other parts of the Circumpolar North.

That said, in general, Greenpeace is a lot freer from its legacy challenges in the Russian North when dealing directly with Indigenous audiences. However, Greenpeace encounters unique state-based challenges associated with its network as a foreign-based non-governmental organization (NGO) operating in a country that is not democratic, with political and legal hostilities.

This chapter briefly introduces Greenpeace's efforts at oil- and gas-focused activism in Russia and the organization's work to help protect sacred territory in the Numto area in ally-ship with local Indigenous peoples. This chapter emphasises the legal and political complexities of operating in Russia and the state's history of environmental problems related to the oil and gas industries and Soviet-era infrastructure. In particular, Chapter 6 stresses the reoccurring theme in *Greenpeace in the Circumpolar North*, which is that Greenpeace appears to be at its best in terms of building fruitful, long-term connections in the North when its agenda both

DOI: 10.4324/9781003674023-6

aligns with local concerns and it involves itself in local issues at the behest of, and in ally-ship with, on-the-ground stakeholders and rightsholders.

The Arctic 30 and the Russian Oil and Gas Industry

Greenpeace's history of campaigning in Russia is complicated by Russia not really being a democracy, with quite a different legal and political history from the Western Arctic states.[3] In 2013, Greenpeace came up against the Russian state when its activists sailed the *Arctic Sunrise*[4] vessel to protest offshore oil extraction (Stewart 2015). The Arctic 30, as it became known, included 28 Greenpeace activists and two freelance journalists who were arrested by Russian authorities and spent two months in jail after a direct action stunt which included protesters scaling "a Russian oil drilling platform in the Pechora Sea" (CBC News 2014; also see Stewart 2015). The fallout of the Arctic 30 protest was a stark surprise for Greenpeace and the Arctic 30 members, who were first arrested on charges of piracy, which was later changed to hooliganism (Dodds and Nuttall 2016, 45–46).

The fallout of the Arctic 30 arrests impacts how Greenpeace now approaches campaigning in Russia. Faiza Outahsen of Greenpeace Netherlands was a member of the Arctic 30. Outahsen reflected that the severity of the Russian authority reaction to the direct action protests was extreme compared to how Greenpeace is typically responded to in the other Arctic states, such as Canada, Kalaallit Nunaat/Greenland, and Norway. In the Western Arctic states, Greenpeace protesters have previously "sought to board [offshore] oil platforms" and had run into the authorities, but the situations did not escalate (Dodds and Nuttall 2016, 45; also see Matz-Lück et al. 2022). Outahsen stated,

> I think if there is one office that has been majorly impacted by this it is the Russian office…Things have drastically changed in Russia over the past years. It doesn't just have to do with this action. It has to do with what happened in Ukraine and how Russia is. I think the Russian government is deeply concerned with protests around the borders. And what happened to us showed the beginning of a new time, you know. I remember, I had one of the staff members of the Dutch consulate visiting me and he lives in Russia, not anymore, I remember that he said to me that his wife who is Russian said "I'm reading stuff in the newspapers that I haven't read since the communist era and this was when we were in prison". So times were changing. That wasn't purely because of our action, I don't think so. There were bigger things going on but we did the action during a time in which things started to change. And I think things are very different in Russia now.[5]

To safely navigate environmental activism in Russia, relatively speaking, Greenpeace needs to lean more on local knowledge and connections in the Russian political and legal ecosystem to tread more cautiously than it might otherwise in other parts of the Circumpolar North.

Ultimately, the Arctic 30 were released from a Russian jail after months of imprisonment. Their arrests happened in the lead-up to Russia's hosting of the 2014 Olympics in Sochi (Olympics 2024), and it brought a lot of negative international attention to Russia at a time when it was actively seeking positive international media coverage. It is likely, therefore, that the Russian government wanted the negative publicity to go away and not interfere in the coverage of the Olympics. Outahsen surmised,

> I think the international pressure helped. We were 30 people with 18 different nationalities. That made a big difference....And the Olympics were coming up. I think that also helped. Again, I'll never be able to give the true answer; nobody knows the real answer unless you're working for the Kremlin.[6]

Ben Stewart was a journalist who was reportedly part of the campaign to free the Arctic 30. Stewart (2015, 172) echoed Outahsen that the timing of the then-upcoming Olympics played a role in helping to get the Greenpeace protesters freed. Diplomats were reportedly advising Greenpeace that "the Sochi Games are Putin's baby" and he was very protective that they would happen with minimal negativity surrounding them.[7] Greenpeace's international coalition of activists from its various national offices formed and ran the Arctic 30 campaign and had a harrowing experience, but Greenpeace's Russian-based members appear to be a bit more contextually in tune with day-to-day cultural, historical, and political factors relevant for legal and governmental engagement in Russia.[8]

Elena Sakirko of Greenpeace Russia recalled that as a result of the Arctic 30 fallout, Greenpeace is more hesitant to use direct action in Russia.

> Sometimes it's not suitable in the Russian context, but we use it [direct action] sometimes, but more before the Arctic 30. After Arctic 30, it [direct action] is connected to a lot of risks for the people working in our organization.[9]

Sakirko went on to say that in her view: "It is very tricky also because some of [locals working with Greenpeace] are being threatened and receive different difficulties at work or lose work because of being [connected] with us campaigning".[10]

At the same time, however, Greenpeace's ability to foster positive working relationships with Indigenous peoples, including in the Circumpolar North, appears to be easier, at least from the perspective of

Vladimir Chuprov of Greenpeace Russia. Chuprov reflected that "Greenpeace Russia has an advantage by having very good relation experience with Indigenous groups, especially in the North and the Far East, much better than other [offices] like the Canadians or the United States for example".[11] This positive position is in part the result of Greenpeace's anti-sealing and anti-whaling legacy from the 1970s and 1980s not having much direct impact or permeability in Russia, as it was part of the Soviet Union when these campaigns occurred. As a result, local peoples were largely insulated from awareness of these Greenpeace activities when they were current news and the impacts of these activists. As Sakirko noted, "Inuit in Russia…we have never had such conflict as [other Greenpeace offices] have had in Canada and Greenland". At the time of the interview, Sakirko did note that Inuit in Russia are somewhat "suspicious" of working with Greenpeace and awareness of the experiences of their colleagues in Inuit homelands in other parts of the North; "because of the work in the other countries…they were not willing to campaign together with us. But we still share information".[12] While this claim of information sharing may be correct, in researching this book, the author was not able to verify this claim of information sharing from a Inuit in Russia official source.

As the Arctic 30 example illustrates, Greenpeace Russia has to operate in a national geopolitical space quite different from the other Greenpeace offices in Arctic countries. Greenpeace Russia must navigate a legal and political system that threatens to label the organization and its members as foreign agents, an ever-present risk for NGOs operating in Russia.[13] Despite this, Greenpeace's Russian office is very active in trying to address issues in the Russian North, most notably in cooperation with Indigenous peoples who have serious concerns over the impact of the oil and gas industry on their homelands and peoples while reportedly maintaining a constructive working dialogue with the Russian government,[14] according to Chuprov.[15]

The oil and gas industry is a double-edged sword for Russia and the Russian North. On the one hand, the industry has been instrumental in the growth of Russia's economy and its resurgence in international politics since the collapse of the Soviet Union. According to a report by the Organisation for Economic Co-operation and Development (OECD), Russia "is the third largest producer of fossil fuels in the world [and] holds [the] second largest proven reserves of natural gas" (OECD n.d.). Capitalizing on its natural resource wealth, Russia has rapidly increased the country's wealth but Sabitova and Shavaleyeva (2015, 427) argue that now "the Russian Federation [is] highly depend on oil and gas revenues" with "oil and gas provid[ing] 39% of the federal budget revenue and [making] up 60% of Russian exports in 2019" (Davydova 2021; also see Warsaw Institute 2020).

On the other hand, the industry has had a negative environmental impact on the Russian wilderness and ecosystems. On May 29, 2020, on the Taymyr Peninsula in the Russian Arctic, Russia experienced the largest known oil spill in the world (Vikulova 2021). News reports state that Russian authorities believe that approximately "21,000 tons of oil poured into the surrounding ground and waterways near the city of Norilsk after a [oil spill from a] diesel oil tank belonging to a subsidiary of Russian metals giant Nornickel" (The Moscow Times 2020). The spill "contaminated the Ambarnaya river and surrounding subsoil" (BBC News 2020). The spill happened because of damage to a storage tank near Norilsk, which investigators suspect "sank because of melting permafrost, which weakened its supports" (BBC News 2020). As a result of the spill, the Ambarnaya River "ran red" due to "a layer of petrochemicals" that coursed through it (Vikulova 2021).

The Norilsk oil spill resulted in a massive wave of public concern and awareness in Russia. According to Greenpeace International:

> Following public pressure, media attention, evidence from Greenpeace Russia — and finally a court battle with the government — the influential company responsible, Norilsk Nickel, has been ordered to pay full compensation for damage caused to the vulnerable Arctic ecosystem. Totalling almost 150 billion rubles (nearly 2 billion USD), it's the largest compensation for environmental damage in Russia's history.
>
> (Vikulova 2021)

The court ruling was a landmark in Russia due to its size and the fact that, in the past, many oil spills in Russia, admittedly on a smaller scale compared to the Norilsk spill, allegedly went unreported and were not legally pursued for clean-up and compensation.[16]

Numto and Working with Local Indigenous Peoples in Russia

Compounding the issue of oil spills is the encroachment of the oil and gas industry into areas previously acknowledged and designated as nature reserves. Arguably, one of the highest-profile examples of industrial creeping into protected areas is the example of Numto Nature Park/reserve. Numto Nature Park was established in 1997, and its size is "more than half a million hectares (about 7218 square km)" (Tysiachniouk and Olimpieva 2019, 60). It is located

> in western Siberia [and] contains a sacred lake, endangered cranes and valuable wetlands for the Nenet and Khanty peoples. The area was also the site of an Indigenous rebellion against the Soviet state during the 1930s, in protest at forced collectivization.[17]
>
> (Kaljur 2017)

Numto's nature reserve protection is under threat after "the regional government re-zoned the nature reserve around the holy Lake Numto, a habitat of the endangered Siberian crane, to allow oil drilling in these wetlands, despite objections from local indigenous residents" (Luhn 2017).[18] The oil company Surgutneftegaz, for example, sought to expand into the Numto Nature Park resulting in the park becoming "an arena of conflict between indigenous people, mostly Forest Nenets who live in the park, and the oil company Surgutneftegaz" (Tysiachniouk and Olimpieva 2019, 57).[19]

In the case of Numto, Greenpeace Russia reportedly became involved in the issue after an activist based near the area approached the organization for help with the Surgutneftegaz encroachment into the Numto Lake area. Specifically, the local activist sought help from Greenpeace to raise local awareness about upcoming, but poorly advertised, public hearings with government officials about the proposed Numto oil exploration.[20] Initially, Greenpeace Russia members reportedly went to the area

> To just talk with people about hearings that were going to happen and that they should participate and speak about what they don't want if they don't want [the new exploration]. They were all very worried about the new exploration. [We talked about] that they should come [to the hearings] and what they should say and we also prepared our analysis of the project and all the risks associated with oil exploration that are unacceptable.[21]

After the hearings ended, Greenpeace did two key things that helped to galvanize the issue and local participation. First, it did a couple of interviews with local peoples and then it also started a petition to stop the Numto plans. In 2016, Greenpeace "organized a worldwide campaign to save Numto Nature Park and challenged the quality of the EIA [Environmental Impact Assessment]" (Tysiachniouk and Olimpieva 2019, 71). The petition "was unexpectedly popular. It was in two weeks we collected more than 36000 signatures, and in Russia. It was only national".[22] Elena Sakirko reflected that the national popularity of Greenpeace Russia's Numto petition was perhaps

> Because it was a sacred place for Indigenous people and very beautiful, unique place and it was a protected area with rare species of birds and wetlands. Scientists also promoted us with some information that is not always traditionally explored.[23]

Sadly, from the perspective of Greenpeace Russia and the local Indigenous peoples it was aiding, they were unable to stop the Numto development plans.

> Unfortunately, in spite of all the letters from local people, from scientists, public hearings where people said they were against [the oil exploration] but still the license, the drilling, was approved on this land, the wetlands. For that they had to change the zone. Before it was a specially protected area, but they changed it saying the scientific analysis [on which the decision to protect the area was made] was old so we need to make a new one and they opened this area to oil exploration but they stepped away from the lake a bit because of the public opinion, because of the sacred lake, but they are still doing exploration on the very fragile area, the wetlands, on which there are a lot of rare species of birds and also quite an important area for Indigenous reindeer herding. As far as after that, they [the authorities] made the area very difficult to reach for us. I know that a couple of journalists who wanted to go there, did not succeed together and were followed by state journalists who explained who they could speak to and who they could not.[24]

Not all of Greenpeace's, or Greenpeace Russia's in this case, campaigns are successful, even with Indigenous participation, invitation and support.

Tysiachniouk and Olimpieva (2019, 71) stated regarding the proposed oil extraction expansion into the Numto area that the economic development plan of Surgutneftegaz progressed despite local concerns "with no mechanism for indigenous people to voice concerns or influence the process".

> Despite sympathizing with the grievances of the indigenous people, the national park's administration depends on Surgutneftegaz's money for park maintenance and infrastructure, so they do not actively support indigenous protests. As a result, dialogue between indigenous stakeholders and the company has become a mere formality.
>
> (Tysiachniouk and Olimpieva 2019, 71)

The company instead allegedly opted to pay off certain residents as it was "easier and economically advantageous [for them]…than to develop complicated mechanisms that would allow [Indigenous residents] to participate and provide them with opportunities to temper the ecological and cultural risks involved".

However, Greenpeace Russia's ability to foster positive relationships with Indigenous communities in the North may be aided by the fact that Vladimir Chuprov is Komi, and he is a long-standing, established member of Greenpeace Russia working in a leadership position.[25] Chuprov reflected that in his experience,

> If I come to a northern community as [a member of] Greenpeace, it helps. In the majority, it overwhelmingly helps. If I come to an Indigenous group and I come to a local community and say I am

> Greenpeace and I am Indigenous it helps twice. Much more, double. Greenpeace is accepted in public opinion as more, a better organisation because it resists the government. It's almost like a joint enemy; the enemy of my enemy is my friend.[26]

Chuprov is an asset to Greenpeace as a subject expert and seasoned activist with networks and insights into how to navigate the overt and unspoken norms and expectations of operating as an activist and NGO in Russia. However, Chuprov is also an asset because of his personal experiences as an Indigenous person raised and living in the Russian Federation that aid his ability to help Greenpeace relate to, communicate with, and navigate expectations of Russian Indigenous peoples with whom the organization may have dialogue or alliances, as well as the realities of dealing with Russian authorities and corporations.

Greenpeace's own appeal as an ally to local peoples, Indigenous or non-Indigenous, is further reinforced in Russia by the organization's networks, skillsets, and resources. After Greenpeace Russia's efforts to aid Indigenous peoples with the protection of the Numto nature reserve, Sakirko reflected that

> Many indigenous peoples reached out to me and my colleagues asking also for help and understanding that Greenpeace is really trying to help local communities and expose violations. So sometimes for some of the people living in the provinces they see the work we do they really understand and like Greenpeace, but some of them are afraid of doing things with us because of pressures, but some of them not.[27]

Perhaps the hope of some Indigenous peoples reaching out to Greenpeace Russia is that the organization can help enable them to bring attention to local issues that people may lack the resources, networks or know-how to amplify in ways that might help get problems, like oil spills, resolved.

According to Sakirko, for example, being able to get on the ground in Russia is a big part of the organization's success. Greenpeace Russia's work includes

> Field trips to different regions, trying to work closely with some of the local communities to help them to tell their stories and to have some joint field checks on what is happening in their territory. We…organised a number of skill shares for local people so they are able to, themselves, report violations because often it is a problem in those regions there can be difficulties to enter Russia.[28]

During reflections on how she became involved with Greenpeace Russia, Sakirko also spoke at length about how the organization has a good track record in Russia, in their opinion, of helping to bring together local groups looking to amplify awareness on environmental issues.

Specifically, Sakirko pointed out that Greenpeace can help local peoples who would like decision-makers to resolve or mitigate environmental damage, but need help getting decision-makers to be aware of, and motivated to, assist them.

> My first project as a staff member was oil spills in the Komi republic where we invited volunteers, journalists and tried to clean up on oil spills in the region [because] it is suffering a lot from oil spills to show how difficult it is to follow the oil chain of how far this oil spill goes. Also we looked, trying to use space images, to detect all the oil spills in the region and then try to make spill checks and report them to the government. So here in two weeks we detected more than 200 oil spills just in the Komi republic...One of the side projects of this oil spill project was to do a local survey with the local population on how oil companies coming to the region have influenced the lives of the Komi people in the villages; how it has changed.... The results of the survey were also astonishing. 95% of all the villages have no gas, the roads are very bad, there is no internet connection and it is one of the riches regions in gas and oil at the same time. The people get nothing actually from this exploration. They are just having polluted water, cannot fish, and things like that.[29]

As the case of the Norilsk oil spill previously mentioned illustrates, Greenpeace Russia's approach to environmental protection appears to be, at least in part, based on trying to help locals near spill sites to address their capacity gap that makes them disproportionately vulnerable to large companies. That being said, the issue of local peoples being cut out of decision-making and the profitability of resource extraction in their areas remains a major issue in Russia.

The issue of locals not being represented in discussions about the use of their homelands[30] or being compensated when environmental damage happens in their local areas was elaborated upon by Sakirko. She recalled a situation in 2016 in which Greenpeace Russia received "a request from a Yamal family that was writing that they have a lot of oil spills and would we [Greenpeace] please come and participate, as in fix them or help us to report them" in the Yamalo-Nenets Autonomous District in Russia.[31] In the 2016 Yamal campaigning (Greenpeace 2016), Greenpeace Russia members went to the closed territory with the permission of local peoples and once inside of the closed territory Greenpeace Russia members claim, with the aid of a videographer, that they helped the Yamal family to report the oil spills after spending "a week there and found around 60 oil spills in the area, not even going deep into the forest".[32]

After Greenpeace Russia reported the oil spills, the issue allegedly morphed into

> A scandal ... because they [Russian authorities] didn't know Greenpeace could go there and were really shocked...It's not always possible [to travel to certain areas] unfortunately. In some areas we could not go, as in physically they [the authorities] just would not let us go.[33]

Sakirko noted that the scandal resulted in the company whose pipeline was responsible for the oil spills being:

> Obliged to pay a big fine and some people were even fired from the company.[34] They did cleanup work in the area, but of course the problem is that the community [near the oil spills] did not get any compensation. The fine money was paid to the government.[35]

The fact that fines are being issued in some instances of oil spills is progress in Russia, and Greenpeace appears to be part of that process of using the media and public pressure to help persuade Russian authorities to consider the merits of being stronger on environmental damage caused by corporate malfeasance in the maintenance of their infrastructure. That said, it can be "hard to assess the impact of a particular campaign or of a particular environmental NGO...Pressure groups may be inclined to exaggerate the degree of influence they have been able to exert" (Newell and Grant 2000, 240). Newell and Grant's assertion that exaggeration occurs in environmental activism may be occurring in Russia, but for her part, Sakirko commented that an overarching reason why Greenpeace Russia is called upon for help by local peoples, particularly local Indigenous peoples, is that the processes for resource development in Russia do an inadequate job of factoring Indigenous peoples into the governance equation.[36]

Russian authorities appear to proceed with decision-making when it relates to oil development as though the land is empty. In particular, Sakirko noted that

> The problem is that the oil companies got licences to explore and it is only after that that they go and finding out that there is Indigenous peoples there in these territories. Of course they know this but there is no legislative mechanism that would consult Indigenous peoples or take into consideration the area in which they are living at the moment when the licence is being issued to the oil companies. That creates a lot of conflicts and confusion because it means that the oil company already has an obligation to explore this land and take oil from the region and it is an area where for centuries, for example, nomadic reindeer herders have been living.[37]

Therefore, even with Greenpeace Russia's efforts to help Russian Indigenous peoples deal with the environmental damage of resource exploration, there remain serious issues with how resource governance and environmental protection are managed in Russia. A big part of the problem in Russia appears to be the exclusion of Indigenous peoples from much of the decision-making processes surrounding the development planning, licensing, benefits sharing, and more of their homelands.[38]

Conclusion

As this chapter illustrates, Greenpeace's ability to build inroads into the North has flourished in instances where Greenpeace was in effect invited to work on local issues by local peoples, predominantly Indigenous peoples. In Finland and Canada, Greenpeace branches were able to successfully support local actors and their agendas to protect ecosystems and combat corporate interests. As illustrated in Chapter 5 and the case of Kangiqtugaapik/Clyde River, Greenpeace was able to help the community affect national law.

In Russia, however, Greenpeace's successes are more tempered, but important, when it comes to serious issues of environmental degradation and remediation in remote areas. Notably, however, the approach to collaboration and campaigning as outlined in the examples in this book, they contrast starkly from Greenpeace's campaigning efforts during the 1970s, most especially on the anti-sealing cause; a complex period in the organization's history that continues to shadow its legacy in environmental conservationism (Woods 1986; Kerr 2014; Burke 2020, 2023, 2024). Though Greenpeace's anti-sealing legacy has minimal salience in Russia, the lessons learned from it are applicable there as well, and as the work in Numto demonstrate, working with locals, demonstrating consistent cultural respect and awareness and taking their campaigning cues from local leaders is proving to be a productive way of building long-term relationships with stakeholders and rightsholders in the Russian North, and regionally as well.

Notes

1 For example, while interviewing Elena Sakirko of Greenpeace Russia, she stated,

> I know, for example, that in … Taimyr they need special permission request for foreigners to enter and it is very difficult to get on. Some of the journalists I was working with were denied this permission. That's why we, as Greenpeace Russian citizens can go but we are followed by FSB and also a lot of the people that we talk to have difficulties, etc. etc. so what we are trying to do now is trying to empower

the local communities to be able to do our work, so to say not really our work but to be able to report violations themselves. It's hard. Some of the people manage to do this but they are the only people that live in this remote land and have interest to report this because often it is the only thing they can do to protect their homeland.

(Interview by Danita Catherine Burke with Elena Sakirko of Greenpeace Russia, September 23, 2018)

2 A similar sentiment was also expressed broadly when I spoke to people from WWF who were also interviewed in 2018–2019 about the challenges associated with doing conservation work in Russia.

3 The other seven Arctic states are Canada, the Kingdom of Denmark (Kalaallit Nunaat/Greenland, the Faroe Islands, and Denmark), Finland, Iceland, Norway, Sweden, and the United States.

4 The *Arctic Sunrise* is a vessel owned by Greenpeace since 1995, which the organization has used extensively as part of its documentation and campaigning about marine oil pollution (Greenpeace International n.d., *Arctic Sunrise*; Malhotra 2024).

5 Interview by Danita Catherine Burke with Faiza Outahsen of Greenpeace Netherlands, September 11, 2018; Outahsen's response was in response to the interview question: "Does the Arctic 30 fallout influence how you approach campaigns involving Russia now?"

6 Interview by Danita Catherine Burke with Faiza Outahsen of Greenpeace Netherlands, September 11, 2018; Outahsen also noted that in her opinion: "The Russians build a case based on total nonsense. That helped a little bit".

7 The Arctic 30 received amnesty from the Russian government as part of a general amnesty bill that was passed in December 2013. Stewart (2015, 331) wrote: "The Kremlin [took] out the trash before the Winter Olympics in Sochi…None of the activists [are] happy that their freedom comes wrapped in Putin's magnanimity, but they [knew] they [wouldn't] get home unless they [took] the amnesty" (also see DW 2013).

8 Interview by Danita Catherine Burke with Elena Sakirko of Greenpeace Russia, September 23, 2018, reflected on the Arctic 30 situation and its impact on how Greenpeace Russia conducts its activities compared to Greenpeace members in the other Arctic states in an interview with Danita Catherine Burke. Burke asked, Greenpeace is notorious in North America for its direct action, which is what it attempted with the Arctic 30, and that didn't go to plan. Does your branch of Greenpeace in Russia ever use direct action in the way that it is typically seen in the media in the west?

9 Interview by Danita Catherine Burke with Elena Sakirko of Greenpeace Russia, September 23, 2018.

10 Interview by Danita Catherine Burke with Elena Sakirko of Greenpeace Russia, September 23, 2018.

11 Interview by Danita Catherine Burke with Vladimir Chuprov, September 27, 2018.

12 Interview by Danita Catherine Burke with Elena Sakirko of Greenpeace Russia, September 23, 2018; Sakirko's comments about the anti-sealing legacy were in response to the question: "With regards to the anti-sealing legacy, do Inuit in Russia have the same relationship to Greenpeace's anti-sealing legacy as do Inuit in Northern Canada and Greenland?"

13 In 2012, Russia started to clamp down on the activities of NGOs with the Foreign Agent Law. According to Martin Russell (2022, 3), the original legislation required organizations getting funding or other property from non-Russian-based sources and who participate in political activities in Russia to declare themselves as foreign agents and "[s]ince June 2014, those that fail to do so can be forcibly registered as such by the Ministry of Justice".

14 Interviews with Greenpeace activists in Russia were conducted before the Russian invasion of Ukraine, and the subsequent fallout for Russia, including sanctions and travel restrictions imposed by Western states (BBC News 2023; Faulconbridge et al. 2023; Nichols 2023) and internal tensions in Russia (Blinken 2023; Council of the EU 2023; Department for Business and Trade 2023).

15 Interview by Danita Catherine Burke with Vladimir Chuprov of Greenpeace Russia, September 27, 2018; Chuprov noted in his interview that he has frequent communication with Russian civil servants and politicians as part of Greenpeace's Russian advocacy work. According to J.P. Richards and J. Heard (2005, 25–26),

> Environmental groups have employed one of two broad approaches to campaigning: they have used either "insider" or "outsider" strategies…Groups that achieve "insider" status are regarded by the government and the public as legitimate, and are in regular consultation with the government.

While Greenpeace is not an insider group per se, it has adopted an increasingly hybrid – insider/outsider – approach to global advocacy, which is notable in the Russian context.

16 Interview by Danita Catherine Burke with Elena Sakirko of Greenpeace Russia, September 23, 2018.

17 The Nenet and Khanty are Indigenous peoples of the Yamalo-Nenets Autonomous Okrug (District) in Northwestern Siberia, Russia (Federation Council of the Federal Assembly of the Russian Federation n.d.; Laptander 2016). Numto Nature Park is located in the Khanty-Mansy Autonomous Okrug, and "is situated in the Beloyarsky district of KhMAO on the border of the Yamalo- NAO in the centre of the West Siberian Plain" (Tysiachniouk and Olimpieva 2019, 60). The Khanti people live primarily in the western area of the park, whereas the Nenets live in the eastern area (Tysiachniouk and Olimpieva 2019, 61).

18 Tysiachniouk and Olimpieva (2019, 61) state that within the park is the village of Numto, a small village of 212 people from which the park takes its name, and it is "located near the sacred Lake Numto.

The lake is worshiped by local people as a 'living god', and the village's main religious ceremonies take place there".

19 While oil extraction activities already occur in parts of the park area, Numto's "economy is traditional and informal. Reindeer herding remains the basis of household economic activity" with the village being "the central reindeer herding settlement in the park" (Tysiachniouk and Olimpieva 2019, 61).

20 Greenpeace Russia has worked on both the wider issues of oil spills and clean-up efforts and the specific issue of the proposed exploration near Numto. In both instances, Greenpeace Russia's relationships and dialogue with local Indigenous peoples are very important to the organization's issue selection and campaigning, according to both Elena Sakirko and Vladimir Chuprov of Greenpeace Russia.

21 Interview by Danita Catherine Burke with Elena Sakirko of Greenpeace Russia, September 23, 2018.

22 Interview by Danita Catherine Burke with Elena Sakirko of Greenpeace Russia, September 23, 2018.

23 Interview by Danita Catherine Burke with Elena Sakirko of Greenpeace Russia, September 23, 2018.

24 Interview by Danita Catherine Burke with Elena Sakirko of Greenpeace Russia, September 23, 2018.

25 Chuprov stated in his interview,

> I have a Bachelor's in the Environment in Ecology. I started working for Greenpeace in 1993. I remember Greenpeace without computers….I was a good time, without cellphones, etc. I had strong motivation to join environmental movements and I considered Greenpeace because it had the position that fulfilled my internal feeling regarding my internal ethical scale. I wanted to join Greenpeace since I was a teenager.
>
> (Interview by Danita Catherine Burke with Vladimir Chuprov of Greenpeace Russia, September 27, 2018)

26 Interview by Danita Catherine Burke with Vladimir Chuprov of Greenpeace Russia, September 27, 2018.

27 Interview by Danita Catherine Burke with Elena Sakirko of Greenpeace Russia, September 23, 2018.

28 Interview by Danita Catherine Burke with Elena Sakirko of Greenpeace Russia, September 23, 2018.

29 Interview by Danita Catherine Burke with Elena Sakirko of Greenpeace Russia, September 23, 2018.

30 The issue of local peoples on the Yamal Peninsula feeling disenfranchised goes back generations due to limited governance controls or opportunities for genuine input into development planning and approvals and infrastructure construction in Siberia. During the Soviet era, "the local councils (*sovet*) have limited authority… Local authorities generally find themselves in an awkward position…[local officials] feel obligated to sign papers they do not like because they are psychologically not yet accustomed to refusing" (Vitebsky 1990, 23).

31 The Yamal peninsula in Western Siberia has a long-standing role in the Russian economy and oil and gas industry, dating back to the Soviet era. At the time that the Soviet Union began to crumble, the Yamal peninsula was the Union's "main producer of gas and oil" with Western Siberia being considered the largest storehouse of hydrocarbons in the world in 1989 (Vitebsky 1990, 19).

32 Interview by Danita Catherine Burke with Elena Sakirko of Greenpeace Russia, September 23, 2018.

33 Interview by Danita Catherine Burke with Elena Sakirko of Greenpeace Russia, September 23, 2018.

34 Interview by Danita Catherine Burke with Elena Sakirko of Greenpeace Russia, September 23, 2018.

35 Sakirko went on to allege that

> The company was kind of punished but the people, because there is no good regulation, they are still reporting things and they want to go to court but it is very difficult because they have no "proper" proof. Their story is that their reindeer died because of the oil spills happened close to their houses but they can't prove it because it is very difficult to prove that they died because of that [directly as a result of the oil spills]. Also all the local specialists involved, they are often working very closely with the oil companies and are not interested in doing independent work. Still it was a situation in which they [the local people] contacted us and we helped them out and this trip brought out other contacts and activists and we are still doing some things together on sharing information and providing some help with their requests.
>
> (Interview by Danita Catherine Burke with Elena Sakirko of Greenpeace Russia, September 23, 2018)

36 Interview by Danita Catherine Burke with Elena Sakirko of Greenpeace Russia, September 23, 2018.

37 Interview by Danita Catherine Burke with Elena Sakirko of Greenpeace Russia, September 23, 2018.

38 Sakirko observed that

> They are trying to find a way to have social-economic agreements, for example in Siberia, but it is also not about saying no to oil exploration but it is just about the financial conditions which means that actually no reindeer herding in the long perspective because you can't do it on the same land as the things [oil exploration]; it's completely impossible. While for the nomadic people it is a bit better with the oil and gas exploration but it is still bad because they lose their routes, they cannot follow the same routes, the territory of Indigenous peoples is getting less and they are getting too many reindeer on the very small area – not very small in our understanding since it is a lot of kilometers but for the reindeer herding and the amount of reindeer that they have its small and not enough. Also, the construction of infrastructure – roads, electricity, and all the

infrastructure that goes with oil exploration – it also creates a lot of difficulties for the Indigenous peoples. Sometimes it brings civilization to some of them, but it ruins their old culture styles like traditional culture styles which is very much based on nature resources: forest, tundra, fish, hunting.

(Interview by Danita Catherine Burke with Elena Sakirko of Greenpeace Russia, September 23, 2018)

References

BBC News. 2020. "Russian Arctic Oil Spill Pollutes Big Lake Near Norilsk." June 9. Available at: https://www.bbc.co.uk/news/world-europe-52977740

BBC News. 2023. "What Is Russia's Wagner Group, and Where Have Its Fighters Gone?" July 18. https://www.bbc.com/news/world-60947877

Blinken, Antony J. 2023. "United States Imposes Additional Sanctions and Export Controls on Russia in Coordination with International Partners." *U.S. Department of State*, May 19. https://www.state.gov/united-states-imposes-additional-sanctions-and-export-controls-on-russia-in-coordination-with-international-partners/

Burke, Danita Catherine. 2020. "Re-Establishing Legitimacy after Stigmatization: Greenpeace in the North American North." *Polar Record* 56(e26): 1–12.

Burke, Danita Catherine. 2022. *WWF and Arctic Environmentalism: Conservationism and the ENGO in the Circumpolar North*. Manchester University Press.

Burke, Danita Catherine. 2023. *Cultural Violence, Stigma and the Legacy of the Anti-Sealing Movement*. Routledge.

Burke, Danita Catherine. 2024. "Nonviolence and Greenpeace: Apologies and Anti-Sealing Campaigning." *Arctic* 77(3): 246–260.

CBC News. 2014. "Two Canadian Members of the Arctic 30 Relive Their Experience in a Russian Jail." *The Current*, January 2. https://www.cbc.ca/radio/thecurrent/jan-2-2014-1.2908860/two-canadian-members-of-the-arctic-30-relive-their-experience-in-a-russian-jail-1.2908868

Council of the EU. 2023. "Russia's War of Aggression against Ukraine: EU Adopts 11th Package of Economic and Individual Sanctions." *European Council*, June 23. https://www.consilium.europa.eu/en/press/press-releases/2023/06/23/russia-s-war-of-aggression-against-ukraine-eu-adopts-11th-package-of-economic-and-individual-sanctions/

Davydova, Angelina. 2021. "Will Russia Ever Leave Fossil Fuels Behind?" *BBC*, November 24. https://www.bbc.com/future/article/20211115-climate-change-can-russia-leave-fossil-fuels-behind

Department for Business and Trade. 2023. "NTE 2023/08: Russia Sanctions – Trade Sanctions Circumvention." *Government of the United Kingdom*, May 22. https://www.gov.uk/government/publications/notice-to-exporters-202308-russia-sanctions-trade-sanctions-circumvention/nte-202308-russia-sanctions-trade-sanctions-circumvention

Dodds, Klaus and Mark Nuttall. 2016. *The Scramble for the Poles: The Geopolitics of the Arctic and Antarctic*. Polity Press.

DW. 2013. "'Arctic 30' Charges Dropped." December 25. https://www.dw.com/en/criminal-charges-dropped-against-29-of-the-arctic-30-greenpeace-says/a-17324108

Faulconbridge, Guy, Darya Korsunskaya and Gleb Stolyarov. 2023. "Russian Elites Trembled as Wagner Rebellion Took Hold." *Reuters*, June 29. https://www.reuters.com/world/europe/russia-teetered-elite-trembled-some-private-jets-left-2023-06-28/

Federation Council of the Federal Assembly of the Russian Federation. n.d. "Yamal-Nenets Autonomous Area." http://council.gov.ru/en/structure/regions/YAN/

Greenpeace. 2016. "Local Activists Oppose Oil Industry in the Yamal District in Russia." July 20. https://media.greenpeace.org/archive/Local-Activists-Find-Broken-Pipeline-in-the-Yamal-District-in-Russia---Web-Video--Russian-Version--27MZIFJJ8MLUB.html

Greenpeace International. n.d.. "The Arctic Sunrise." https://www.greenpeace.org/international/about/ships/arctic-sunrise/

Kaljur, Lauren. 2017. "Why Russia's Indigenous People are Wary of National Parks." *The New Humanitarian*, May 9. https://deeply.thenewhumanitarian.org/arctic/articles/2017/05/09/why-russias-indigenous-people-are-wary-of-national-parks

Kerr, Joanna. 2014. "Greenpeace Apology to Inuit for Impacts of Seal Campaign." *Greenpeace*, 24 June. https://www.greenpeace.org/canada/en/story/5473/greenpeace-apology-to-inuit-for-impacts-of-seal-campaign/

Laptander, Roza Ivanovna. 2016. "Nenets, Khanty and Selkup: The Nenets, Khanty and Selkup Language in Education in the Yamal Region in Russia. Regional Dossiers Series." Mercator European Research Centre on Multilingualism and Language Learning. ISSN: ISSN-1570-1239. https://eric.ed.gov/?id=ED573598

Luhn, Alec. 2017. "The Reindeer Herder Struggling to Take on Oil Excavators in Siberia." *The Guardian*, March 17. https://www.theguardian.com/world/2017/mar/17/reindeer-herder-oil-excavators-siberia

Malhotra, Sudhanshu. 2024. "Best of Greenpeace Ships 2024." *Greenpeace International*, December 18. https://www.greenpeace.org/international/story/71926/best-of-greenpeace-ships-2024/

Matz-Lück, Nele, Øystein Jensen, and Elise Johansen. 2022. *The Law of the Sea: Normative Context and Interactions with Other Legal Regimes*. Routledge Search on the Law of the Sea Series. Routledge.

Newell, Peter and Wyn Grant. 2000. "Environmental NGOs and EU Environmental Law." In *Yearbook of European Environmental Law*, Volume 1. Edited by Han Somsen. Oxford University Press.

Nichols, Tom. 2023. "The Coup Is Over, but Putin Is in Trouble." *The Atlantic*, June 24. https://www.theatlantic.com/international/archive/2023/06/putin-russian-coup-over-prigozhin-wagner/674522/

OECD. n.d. "Russian Federation." https://stats.oecd.org/fileview2.aspx?IDFile=09aac246-c7ef-4159-898e-2a287deb3341

Olympics. 2024. "Olympic Winter Games Sochi 2024." https://olympics.com/en/olympic-games/sochi-2014

Richards, J.P. and J. Heard. 2005. "European Environmental NGOs: Issues, Resources and Strategies in Marine Campaigns." *Environmental Politics* 14(1): 23–41.

Russell, Martin. 2022. "'Foreign Agents' and 'Undesirables': Russian Civil Society in Danger of Extinction?" *European Parliamentary Research Service*. https://www.europarl.europa.eu/thinktank/en/document/EPRS_BRI(2022)729297

Sabitova, Nadia and Chulpan Shavaleyeva. 2015. "Oil and Gas Revenues of the Russian Federation: Trends and Prospects." *Procedia Economics and Finance* 27(2015): 423–428.

Stewart, Ben. 2015. *Don't Trust, Don't Fear, Don't Beg: 100 Days as a Prisoner of Putin – The Story of the Arctic 30*. Guardian Books and Faber & Faber Ltd.

The Moscow Times. 2020. "Russia Admits to 'World's Largest' Arctic Oil Spill." December 28. https://www.themoscowtimes.com/2020/12/24/russia-admits-to-worlds-largest-arctic-oil-spill-a72477

Tysiachniouk, Maria and Irina Olimpieva. 2019. "Caught between Traditional Ways of Life and Economic Development: Interactions between Indigenous Peoples and an Oil Company in Numto Nature Park." *Arctic Review on Law and Politics* 10(2019): 56–78.

Vikulova, Olesya. 2021. "A Year after the Norilsk Disaster, Where Are Russia's Oil Risks and What Needs to be Done?" *Greenpeace International*, May 29. https://www.greenpeace.org/international/story/47973/norilsk-oil-spill-disaster-russia-accident-risk-map/

Vitebsky, Piers. 1990. "Gas, Environmentalism and Native Anxieties in the Soviet Arctic: The Case of Yamal Peninsula." *Polar Record* 26(156): 19–26.

Warsaw Institute. 2020. "Russia's Economy is Becoming Heavily Dependent on Hydrocarbons." February 24. https://warsawinstitute.org/russias-economy-becoming-heavily-dependent-hydrocarbons/

Woods, S.J. 1986. "The Wolf at the Door: The Anti-Harvest Campaign Strikes at the Heart of North Aboriginal Economies." *Northern Perspectives* 14(1): 1–8.

7 Conclusion

Greenpeace's strength and major stumbling block in its efforts to engage more on Arctic and northern issues in the 21st century is its legacy. As *Greenpeace in the Circumpolar North* illustrates, Greenpeace has long purported that Indigenous peoples and rights are important to the organization, but it has also participated in campaigning, such as its work against sealing (and fur trading more broadly) and whaling, which harmed Indigenous cultures and economies and instilled doubt about the organization's commitment to Indigenous rights and autonomy.

On top of this, Greenpeace formed and defined its brand in an era of peak colonially infused understandings within environmental activism of Indigenous and working-class peoples, and what it means to protect the environment. Greenpeace is embedded in a wider movement of environmentally focused organizations originating in the West and predominately urban-based with connection to trends of colonial thinking, actions, and behaviours, which Greenpeace acknowledged to some extent in its 2014 apology to Inuit, and Indigenous and coastal peoples of Canada for features of its and other activists' anti-sealing activism. Though Greenpeace is taking some steps to bridge the gap between its origins and aspects of its campaigning legacies that create barriers to its resonance with various peoples in the Circumpolar North, it still struggles to move past some antiquated aspects of its understanding of peoples, places, and cultures.

The early Save the Arctic work in Kalaallit Nunaat/Greenland in 2010–2011, for example, was driven by outside-the-region factors, namely, the Deepwater Horizon incident in the Gulf of Mexico. While the aim to prevent such an accident from occurring in the Arctic is laudable, the approach taken did not adequately account for local Indigenous, namely, Kalaallit/Greenlandic, considerations when exploring the possibility of offshore oil exploration. The assumption that because someone is Indigenous inherently means that they oppose all forms of resource extraction and wish to leave the environment untouched is grounded in the kind of thinking that led to the development and popularity of the "Crying Indian" public service announcement in the US. In Kalaallit

DOI: 10.4324/9781003674023-7

Nunaat/Greenland, the willingness to consider oil exploration is part of a wider national debate about Kalaallit/Greenlandic self-determination. This opportunity to create a nuanced engagement with Kalaallit/Greenlandic people over oil exploration and avenues for economic independence that do not focus on turning Kalaallit Nunaat/Greenland and Kalaallit/Greenlandic culture and society into a performative venue for tourists, for example, was missed at that time.

However, Greenpeace has taken steps to express and enact change to aspects of its attitudes, beliefs, and awareness of Indigenous rights and the existence and value of Indigenous traditional knowledge, practices, and authority. The organization appears to be trying to illustrate through action, in some parts of the Circumpolar North at least, that it is changing in more than just platitudes, with Indigenous peoples receiving the majority of this consideration.

But Greenpeace's journey to modify its stance on sealing and fur trading was not as smooth as contemporary discussions would present. Rather than stepping up initially and seeking to address errors proactively in how it approached the sealing issue and the impact of anti-sealing and anti-fur trading on Indigenous and non-Indigenous peoples, cultures, and communities, Greenpeace's stance change was reactionary. As Chapter 3 unpacks, it took Indigenous leaders to push Greenpeace to confront some aspects of its part in the harms inflicted on Indigenous communities as a result of anti-sealing and anti-fur activism. From the perspective of the leading Indigenous actors confronting Greenpeace in the mid-1980s, the organization showed reluctance to come to the dialogue table. But it did show up eventually. By late 1986, Greenpeace, at the behest of Indigenous leaders at ISI, withdrew from official anti-sealing and anti-fur campaigning, though there was significant resistance within the organization, especially from within the United Kingdom. This marks the start of a long journey for Greenpeace to reconcile its anti-sealing legacy; a journey that is in need of more consideration of the generational relationship between Indigenous and non-Indigenous sealing traditions, cultures and socio-economic practices and work to address harms to Indigenous who are not Inuit and deeply embedded non-Indigenous rural, working class peoples, cultures and communities.

The look into the three contemporary examples of Greenpeace's efforts to work with Indigenous peoples in the Circumpolar North offers some insight into the importance Greenpeace assigns to improving its relationships with Indigenous peoples. Chapter 4 highlights Greenpeace's work with Sámi reindeer herders in Finland, with land and ecosystem conservation vital to traditional reindeer husbandry. Chapter 5 discusses Greenpeace's effort to assist the Inuit of Kangiqtugaapik/Clyde River, Canada, with their battle to protect treaty rights around consultation as part of the community's fight to protect their local ecosystem from the repercussions of seismic testing/blasting on marine animals. Finally, Chapter 6 unpacks Greenpeace's effort to participate in environmental activism in Russia against oil exploration, the unique challenges of operating in Russia, and its attempts to assist various Indigenous peoples deal with oil spill clean-up and preservation in the Numto Nature Park against encroaching oil industry ambitions.

A common thread in each of these cases is that, according to Greenpeace members involved, the organization only became involved in these examples after local northerners approached the organization for help. This reported approach to northern campaigning is an observable shift from Greenpeace's early activism campaigning against commercial seal hunting in the 1970s–1980s, and whaling too, which had major long-term implications for Indigenous and non-Indigenous rural and coastal fishers throughout the Circumpolar North. I suggest in *Greenpeace in the Circumpolar North* that Greenpeace's modified approach is now more deliberate in its campaign planning and consideration of northern Indigenous peoples, lands, and rights in certain circumstances. This shift signals that the organization is learning and incorporating some lessons from its past campaigning missteps, mistakes, and legacy of harm in the Circumpolar North.

With its anti-sealing activism, Greenpeace activists invited themselves into predominantly rural locations and conducted top-down activism based on the personal preferences of leadership at that time. Topic selection and campaign approaches appear to have been largely driven by what leading personalities in the organization perceived as environmental problems and their personal views and philosophies on appropriate human-nature co-existence. The result was that Greenpeace members imposed their ways of thinking and agendas, promoted misinformation and cultural disregard to national and international audiences, dismissing Indigenous and non-Indigenous traditional knowledge, and largely omitting consistent, respectful, and tolerant local inclusion and leadership in their campaign formulation and strategic planning in their anti-sealing work. In contrast to the circumstances outlined in Chapter 3, there is evidence to suggest, as Chapters 4–6 unpack, that some course correction is occurring to mitigate the likelihood of a repeat of what occurred in the 1970s–1980s.

The organization's capacity to build on its reconciliation for its anti-sealing legacy has so far been curtailed by the fact that many Greenpeace members and supporters remain proud of that legacy and unsupportive of distancing the organization from it. As a result, Greenpeace is performing a bit of a tap dance whereby it is clear that some key members who work in the Circumpolar North and have substantial regional and institutional knowledge about the anti-sealing history, know that this legacy is a barrier to Greenpeace's ability to grow in regional political reach. That said, there appear to be limits on how far Greenpeace can, or would be willing to go, to fully address this history.

Presently, the focus is dominated by discourse and outreach to Indigenous peoples. Indigenous engagement is a good step in addressing the legacy, but it is also one dimension of a broader legacy. But even the Indigenous dimension receives pushback from some Greenpeace members, donors, and supporters.

The 2010–2011 attempt by Greenpeace Nordic, with the main push coming from its Denmark office, to address the anti-sealing legacy in Kalaallit Nunaat/Greenland was an astute attempt at trying to build a bridge between the organization and Inuit in Kalaallit Nunaat/Greenland. This diligent work from Greenpeace Nordic (Denmark) reflects a quiet record of some Danish-based Greenpeace employees being more attuned to the need to reflect on Kalaallit Nunaat/Greenland interests more in the organization's northern work dating back into the 1980s. But, the Greenpeace Nordic inroads were impacted adversely by the wider organizational and supporter concern about Greenpeace appearing to support the killing of seals. The 2014 apology to Canadian-based Inuit, Indigenous, and coastal peoples was a more articulate and planned attempt to navigate the same issues Greenpeace Nordic tried, and appears to have been more outwardly accepted by Greenpeace membership.

The substance of the apology overwhelmingly focused on Inuit and lacks detail about what specifically Greenpeace did (e.g. details of campaigning activities and actions) that warrants an apology to various groups or individuals it impacted. The apology is also light on detail on the other Indigenous and coastal peoples impacted who are not Inuit. Furthermore, the apology gives little detail about next steps, such as what Greenpeace will be doing to find solutions to its past conduct and the persistent repercussions of anti-sealing activism, which Greenpeace helped pioneer and normalize. The apology, however, was timely. It came at just the right time for Kangiqtugaapik/Clyde River (Chapter 5), helping to open dialogue with the community and for them to forge a successful working relationship to stop seismic testing and assert Inuit treaty rights.

That being said, a word of caution is warranted when considering the generalizability of the examples presented in *Greenpeace in the Circumpolar North*. While the examples in Finland, Canada, and Russia do portray Greenpeace in a generally positive light and are arguably good indications of potential organizational growth and learning, it is still too soon to assess to what extent these examples, in conjunction with more recent activities, indicate a true shift in organizational thinking, and consequently, behaviours and actions in the long run. It is possible that these examples illustrate pockets of enlightenment and growth within the organization, concentrated amongst some frontline activists working directly with northerners, but their increasing insightfulness may not have permeated within the organizational culture and personnel at large. Greenpeace, as an organization, may not be able to resist involving itself in local issues without local invitations. This could derail the progress Greenpeace appears to be making in addressing aspects of its negative reputation in the North. But as Mads Flarup Christensen of Greenpeace Nordic (Denmark) reflected when commenting on the campaigns explored in this book: "The question remains to which degree these campaigns are one-offs or [are] something that can be replicated and [will be] informing of a broader picture. That I don't know".[1]

Note

1 Interview by Danita Catherine Burke with Mads Flarup Christensen of Greenpeace Nordic, February 7, 2019.

Appendix A
Background on Primary Book Sources

Greenpeace in the Circumpolar North is the product of an accumulation of research from overlapping projects I have done on environmental and animal rights organizations' involvement in Arctic diplomacy and politics, and the legacy of the anti-sealing movement. As such, the interview and archival research used in this book was supported by European Union's Horizon 2020 research and innovation programme under the Marie Skłodowska-Curie grant agreement No 746312 (2018–2020), a fellowship from the J.R. Smallwood Foundation for Newfoundland and Labrador Studies (2020–2022), a Northern Scholars Visiting Research Fellowship at the Institute for Advanced Studies in the Humanities (IASH), University of Edinburgh (2022–2023), and a Joan Mitchell Travel Award from Wilfrid Laurier University (2025). The finalization of this manuscript occurred while the principal investigator was completing the research project *Seals, Stigma and Survival*, supported by the Nordic Council in Greenland's Nordic Arctic Programme (2025–ongoing).

Archival research included in this book comes from two sources: the Laurier Archives based at Wilfrid Laurier University in Waterloo, Canada, and the Centre for Newfoundland Studies (CNS) based at Memorial University of Newfoundland and Labrador's Queen Elizabeth 2nd Library in St. John's, Canada. The data collected at the CNS was collected as part of the J.R. Smallwood Fellowship in 2020–2022. The CWS holds extensive records on the early years of the anti-sealing movement (1960s–1980s) and the historical importance and history of sealing in the economy and culture of Newfoundland and Labrador, with a focus on the materials accessed during the time period after the turn of the 20th century.

The archival work at the Laurier Archives was conducted in 2025 with the support of a Joan Mitchell Travel Award (awarded in 2023 but deferred due to sick leave for cancer treatment). The Laurier archival holdings include sources on the impact of anti-sealing and anti-fur trading activism and sentiments in the 20th century on northern, predominantly Indigenous, communities, economies, and societies in Canada and

Greenland. These holdings include insight into the persistent work and effort by Indigenous peoples to reach out to and encourage Greenpeace to halt its anti-sealing/fur trade campaigning and barriers that Greenpeace faced in coming to that decision in the 1980s.

The conceptual work on nonviolence was done while a Northern Scholars Visiting Research Fellow 2022–2023. It was at the time that I was a research fellow at IASH, working on this book in spring 2023, that I was diagnosed with cancer and stopped all work for two years on this book project to undergo treatment. This was an unexpected delay in the long process of completing this book, and may therefore affect the timeliness of its conclusions, given changes in socio-economic and geo-political factors in the Circumpolar North and global politics (e.g. progression of the Russian war against Ukraine and its impact on Arctic/northern politics; election of US President Donald Trump).

The majority of the interviews referenced in this book were completed while I was a Marie Skłodowska-Curie fellow (2018–2020), with the exception of the interview with James Winter, which was completed as part of the project supported by the J.R. Smallwood Fellowship (2022–2023). See Table A.1 for more details on the interviews included in *Greenpeace in the Circumpolar North.* These long-form semi-structured interviews were conducted in-person and via video call. The process of identifying and getting in touch with interviewees was initiated with the help of a colleague and former co-author, Teale Phelps Bondaroff, an expert in environmental and animal rights activism and organizational strategies, techniques, and campaigns. Once interviews started, I built upon the contacts using the snowball method of referrals from interviewees and people with whom I had informal conversations as part of the fieldwork process. The snowball method

> may be defined as any type of sample recruitment strategy, whereby all or a portion of participants who are asked to provide data are not directly recruited by the researcher but through other persons who connect them to other persons as participants.
>
> (Marcus et al. 2017, 636)

The interviews with Greenpeace members occurred in the same time period as I was also interviewing members of the World Wide Fund for Nature (WWF); for which more information on the use of those interviews can be found in my 2022 book *WWF and Arctic environmentalism: Conservationism and the ENGO in the Circumpolar North* published by Manchester University Press and my 2020 book chapter "Trust and the WWF Global Arctic Program" in the edited volume *Non-Human Nature in World Politics: Theory and Practice*, edited by Joana Castro Pereira and André Saramago and published by Springer (Burke, 2020; Burke, 2022).

Table A.1 Interviewees Quoted in the Book

Names[a]	*Date*	*Description*
Sune Scheller	Sept 3, 2018	Greenpeace Nordic – Works in Greenland on issues of community outreach and non-renewable resource extraction (based in Denmark)
Mads Flarup Christensen	Feb 7, 2019	Greenpeace Nordic – Executive director of Greenpeace Nordic (based in Denmark)
Sini Harkki	Sept 21, 2018	Greenpeace Nordic – Works with Sámi on protecting reindeer herding and forestry (based in Finland)
Faiza Outahsen	Sept 11, 2018	Greenpeace Netherlands – Was a member of the Arctic 30 and arrested in Russia, protesting Arctic oil extraction
Elena Sakirko	Sept 23, 2018	Greenpeace Russia – Works on issues in the Russian North like oil spill clean-up, protected area preservation, and Indigenous rights, and aided the Arctic 30 as a translator
Vladimir Chuprov	Sept 27, 2018	Greenpeace Russia – Work with Russian northern Indigenous communities and work on Numto protection and oil spill prevention and remediation
Jessica Wilson	Oct 30, 2018	Greenpeace International (formerly of Greenpeace Canada) – Work with Kangiqtugaapik/Clyde River, Save the Arctic team member, and part of the 2014 Greenpeace Canada apology to Canadian Inuit process
John Hocevar	Nov 9, 2018	Greenpeace USA – Global coordination team member for work on the ocean, including Arctic/Alaska work on Shell oil exploration and polar work on underwater canyon protection
Jerry Natanine[b]	Dec 19, 2018	Mayor of Kangiqtugaapik/Clyde River, Nunavut, Canada
James Winter[c]	Dec 28, 2021	Former CBC journalist and founding president of the Canadian Sealers Association

[a] References to participant backgrounds are based on their status at the time of the interviews.

[b] The interview with Jerry Natanine was published in the *Arctic* journal prior to the publication of this book, and it is this interview publication that is referenced in the text.

[c] The interview with James Winter was published on the *Women in the Arctic and Antarctic* initiative website prior to the publication of this book, and it is this interview publication that is referenced in the text.

Furthermore, I also interviewed Indigenous peoples representatives and Arctic state representatives at the Arctic Council, and was working on Arctic-focused regional diplomacy at the same time period. All of these interviews were conducted as part of a project exploring the role of environmental organizations – focusing on the case studies of WWF and

Greenpeace – in Arctic diplomacy. While the WWF and other interviews are not directly quoted in this book, the overarching research in which the interviews with Greenpeace members are part of is reflected in this book. Greenpeace members identified in this book consented to the use of their names in connection to quotes from their interviews.

Two additional interviews were completed with members of Greenpeace – one in 2016 and one in 2018 – but they are not included in this book. The 2016 interview is used in other works that I have written on Greenpeace and the legacy of the anti-sealing movement. These already published sources are quoted in the book. The second interview not used in the book is from a 2018 interviewee who made it a condition for the interview that they be sent quotes in advance for approval before their inclusion into my research. The interviewee's requirement is something that I may have been able to accommodate in other circumstances, but it is not practical here as a result of the time lapse between the interview and the finalization of the *Greenpeace in the Circumpolar North* manuscript. The publication of *Greenpeace in the Circumpolar North* was delayed due to the implications of a career modification necessitated by COVID-19, as well as subsequent long-term sick leave as I underwent cancer treatment.

Additionally, one interviewee, who is named in the book, provided information, used in Chapter 2, on their views on the legacy of Paul Watson in Greenpeace. Though I have permission to attribute quotes to them from their interview in my research, given the specific nature of the quote and the quote's subject matter, I chose not to directly attribute their name out of an abundance of caution for this individual.

References

Burke, Danita Catherine. 2020. "Trust and the WWF Global Arctic Program." In *Non-Human Nature in World Politics: Theory and Practice*, edited by Joana Castro Pereira and André Saramago. Springer.

Burke, Danita Catherine. 2022. *WWF and Arctic Environmentalism: Conservationism and the ENGO in the Circumpolar North*. Manchester University Press.

Marcus, Bernd, Oliver Weigelt, Jane Hergert, Jochen Gurt and Petra Gelléri. 2017. "The Use of Snowball Sampling for Multi Source Organizational Research: Some Cause for Concern." *Personnel Psychology* 70(2017): 635–673.

Index

Pages in **bold** refer to tables and pages followed by "n" refer to notes.